Industrial Process Control:
Advances and Applications

Industrial Process Control: Advances and Applications

Ghodrat Kalani

An Imprint of Elsevier Science

Boston   London   New York   Oxford   Paris   Tokyo
Amsterdam   San Diego   San Francisco   Singapore   Sydney

Butterworth-Heinemann is an imprint of Elsevier Science.

Copyright © 2002, Elsevier Science (USA). All rights reserved.

No part of this publication may be reproduced, stored in a retrieval system, or transmitted in any form or by any means, electronic, mechanical, photocopying, recording, or otherwise, without the prior written permission of the publisher.

Recognizing the importance of preserving what has been written, Elsevier-Science prints its books on acid-free paper whenever possible.

Elsevier Science supports the efforts of American Forests and the Global ReLeaf program in its campaign for the betterment of trees, forests, and our environment.

Library of Congress Cataloging-in-Publication Data
Kalani, Ghodrat.
　Industrial process control : advances and applications / Ghodrat Kalani.
　　p. cm.
　Includes bibliographical references and index.
　ISBN 0-7506-7446-6 (alk. paper)
　　I. Process control.　I. Title.
TS156.8 .K347 2002
670.42′7—dc21
　　　　　　　　　　　　　　　　　　　　　2002071151

British Library Cataloguing-in-Publication Data
A catalogue record for this book is available from the British Library.

The publisher offers special discounts on bulk orders of this book.
For information, please contact:

Manager of Special Sales
Elsevier Science
225 Wildwood Avenue
Woburn, MA 01801-2041
Tel: 781-904-2500
Fax: 781-904-2620

For information on all Butterworth-Heinemann publications available, contact our World Wide Web home page at: http://www.bh.com

10 9 8 7 6 5 4 3 2 1

Transferred to digital printing 2006

Contents

Preface   ix

Acknowledgment   xiii

1 Objectives   1
- 1.1 Introduction   1
- 1.2 Philosophies   1
- 1.3 Requirements   2

2 An Overview of Instrumentation, Control, and Safety Systems   7
- 2.1 Introduction   7
- 2.2 Piping and Instrumentation Diagrams   7
- 2.3 Safety and Automation Systems   9
    - 2.3.1 System Configuration   9
    - 2.3.2 Local Area Network   9
    - 2.3.3 Control Unit   15
    - 2.3.4 Operator Station   16
    - 2.3.5 Host Computer   18
- 2.4 Interface to Foreign Systems   18
    - 2.4.1 General   18
    - 2.4.2 Object Linking and Embedding (OLE) for Process Control   19
- 2.5 Field Instruments   22
    - 2.5.1 General   22
    - 2.5.2 Profibus   23
    - 2.5.3 Foundation Fieldbus   26
- 2.6 Multiphase Flow Metering   33
- 2.7 Subsea Control and Instrumentation Systems   34
    - 2.7.1 Subsea Field Control and Instrumentation   34
    - 2.7.2 Umbilical   37
    - 2.7.3 Master Control Station   37
- 2.8 Vessel Control Systems   38
    - 2.8.1 General   38
    - 2.8.2 Dynamic Positioning   38
    - 2.8.3 Ballast Control   39
    - 2.8.4 Environmental, Meteorological, and Platform Monitoring   40

    2.8.5 Operator Interface 40
 2.9 Condition Monitoring 40
    2.9.1 General 40
    2.9.2 Maintenance Strategies 41
 2.10 Reliability and Availability Analysis 41
    2.10.1 Introduction 41
    2.10.2 Analysis 42
    2.10.3 Standards 46
 2.11 Safety Integrity Level and IEC 61508 47
    2.11.1 Introduction 47
    2.11.2 Risk Analysis 47
    2.11.3 Safety Requirements 49
    2.11.4 Safety Integrity Levels 49
    2.11.5 A Practical Approach 50

3 Systems Theory 53

 3.1 Introduction 53
 3.2 Other Theories 56
    3.2.1 Sampling Theory 56
    3.2.2 Control Theory 58
    3.2.3 Information Theory 60
 3.3 Hierarchical System Configuration 62
 3.4 Open System Technology 64
 3.5 Universal Operating System 68
 3.6 Hardware Variety 69
 3.7 Software Robustness 70

4 Integrated Safety and Automation Systems 73

 4.1 Introduction 73
 4.2 Early Integrated Systems 74
 4.3 Current Integrated Systems 76
 4.4 Future Integrated Systems 76
 4.5 Benefits of Using Integrated Systems 78

5 Project Engineering of Control Systems 81

 5.1 Safety and Automation Systems Configuration 81
    5.1.1 System Hardware 83
    5.1.2 System Software 85
 5.2 Critical Path Methods 86
    5.2.1 Introduction 86
    5.2.2 System Analysis 87
    5.2.3 System Philosophies 87
    5.2.4 System Requirements 87
    5.2.5 Applicable Regulations, Standards, and Codes of Practice 89
    5.2.6 Vendor Selection 93

       5.2.7 Process Interface Design   94
       5.2.8 Operator Interface Design   95
       5.2.9 Staff Training   95
       5.2.10 Application Software Development   96
       5.2.11 System Testing   97
   5.3 Commissioning   98
   5.4 Project Staffing   99
   5.5 Documentation   101

6 Application Engineering of Control Systems   105

   6.1 Introduction   105
   6.2 Typical Control Schemes   106
   6.3 Advanced Control   106
       6.3.1 Level Control by PID Gap   106
       6.3.2 Dynamic Feedforward Control   107
       6.3.3 Heat Balancing Control   108
       6.3.4 pH Control by PID Error Squared   111
       6.3.5 Multivariable Control   111
   6.4 Traditional Control Algorithms   114
       6.4.1 PID Algorithm   115
       6.4.2 Lead/Lag Algorithm   117
       6.4.3 Square Root Algorithm   117
   6.5 Intelligent Systems   118
       6.5.1 Introduction   118
       6.5.2 Artificial Intelligence   119
       6.5.3 Intelligent System Techniques   119
       6.5.4 Neural Networks   121
       6.5.5 Application Example 1   127
       6.5.6 Application Example 2   128
   6.6 Simulation   129
       6.6.1 Introduction   129
       6.6.2 Simulation and Process Control   130
       6.6.3 Simulation and Training   132

7 Some Typical Control and Safety Systems   137

   7.1 Introduction   137
   7.2 ABB Master   137
       7.2.1 An Overview of ABB Master   137
       7.2.2 ABB Master System Internal Communications   138
       7.2.3 ABB Master Operator Interface   141
       7.2.4 ABB Master Control Units   142
       7.2.5 ABB Master Host Computer   143
       7.2.6 ABB Master External Interfaces   143
       7.2.7 ABB Master Application Packages   143
   7.3 Fisher PROVOX   144

        7.3.1 Introduction   144
        7.3.2 PROVOX Communications   145
        7.3.3 PROVOX Operator Interface   146
        7.3.4 PROVOX Control Units   147
        7.3.5 PROVOX Software/Computing Facilities   147
        7.3.6 PROVOX Engineering Tools   148
   7.4 Honeywell TDC 3000   148
        7.4.1 Introduction   148
        7.4.2 TDC 3000 Communications   149
        7.4.3 TDC 3000 Operator Interface   150
        7.4.4 TDC 3000 Control Units   152
        7.4.5 TDC 3000 Host Computer   153
        7.4.6 TDC 3000 External Interfaces   154
        7.4.7 TDC 3000 Application Packages   154
   7.5 Siemens SIMATIC PCS7   154
        7.5.1 Introduction   154
        7.5.2 SIMATIC PCS7 System Overview   155
        7.5.3 SIMATIC PCS7 System Communications   156
        7.5.4 SIMATIC PCS7 Operator Interface   157
        7.5.5 SIMATIC PCS7 Control Units   157
        7.5.6 SIMATIC Supervisory Computer   159
        7.5.7 SIMATIC Engineering Interface   159
   7.6 Silvertech Sentrol   160
        7.6.1 An Overview of Sentrol   160
        7.6.2 Sentrol Internal Communications   161
        7.6.3 Sentrol Operator Interface   161
        7.6.4 Sentrol Control Units   161
        7.6.5 Hazardous Area Applications   163
        7.6.6 Powertools   163
   7.7 Simrad AIM   164
        7.7.1 An Overview of AIM   164
        7.7.2 AIM System Internal Communications   164
        7.7.3 AIM Operator Interface   164
        7.7.4 AIM Control Unit   166
        7.7.5 AIM Host Computer   166
        7.7.6 AIM External Interfaces   166
        7.7.7 AIM Application Packages   167
   7.8 Conclusion   168

Glossary of Words and Abbreviations   171

References   179

Index   181

Preface

Since the publication of my first book[1] in 1988, in which I described the application and project engineering of distribution control systems (DCS), many aspects of control and instrumentation (C&I) systems have undergone major changes. These changes have been caused by an ever-increasing demand to improve productivity, safety, and profitability. Control systems using expert systems, neural networks, IEEE 802 Standard local area networks (LANs), and integrated system architectures are specified by most end-users. Such control systems would not have been possible without the evolution that has taken place during the last decade in electronics, software engineering, and information technology (IT).

There have also been fantastic developments in many aspects of field instrumentation, as follows:

- *Transmitters and Valves.* By using intelligent instruments, significant benefits in calibration/recalibration, rangeability, diagnostics, and reliability of C&I systems have been realized.
- *Fieldbus.* Major reductions in cabling, junction boxes, galvanic isolators, marshalling cabinets, termination assemblies, and space requirements are possible.
- *Multiphase Flow Metering (MPFM).* The use of MPFM instead of test separators in offshore platforms or subsea installations can achieve substantial cost reductions. In some marginal oil/gas fields, the applications of MPFM may render the field viable where otherwise it would have been uneconomical. If MPFM is employed, substantial reductions in initial equipment costs, space, weight, instrumentation, and improvement in productivity are possible.

The present-day C&I systems for plants of medium to large sizes are so powerful and complicated that only a design/application based on the systems theory can ensure success and full system utilization. The suppliers of C&I systems are constantly enhancing the capabilities of their products by employing the latest technological developments; however, not many vendors have exploited the benefits of systems theory application because of a lack of appreciation for the systems theory.

The systems theory will be studied in Chapter 3. It is useful to note that such features as open system architecture, universal operating system, integrated system configuration, information theory, and sampling theory are a direct result of the application of systems theory. I had the opportunity to evaluate more than a dozen control and safety systems for a large offshore oil and gas project in 1992–1993. The systems belonged to suppliers of international repute and various countries (e.g., United Kingdom, United States, Germany, Japan, and Sweden). Although all of the systems were powerful and could satisfy the control and monitoring requirements of the project, only two systems could offer all the features highlighted previously.

The control system specification of requirements that I prepared asked for the aforementioned features. All of the vendors claimed that their systems could meet the specified requirements. During discussions in which the requirements were explained in more detail, however, it became apparent that most vendors do not understand the requirements. Some vendors stated that the features were not necessary for offshore platforms and that their system had been used successfully in similar applications. If end-users accepted such claims, we would still be installing the large case instruments of the 1950s and 1960s with thousands of air tubes around the plants.

When Honeywell introduced the TDC 2000 in 1975, most instrument vendors claimed that the TDC 2000 did not provide single-loop integrity, and consequently was not reliable and suitable for process plants: End-users did not listen to such a biased claim, and Honeywell captured more than 50 percent of the market in the early 1980s. Recognizing the TDC 2000's success, other vendors introduced their microprocessor-based control systems and multiloop control units. At present, I do not know any vendor who mentions or offers single-loop integrity.

This situation equally applies to the present-day control systems with regard to open system technology, universal operating system, integrated system configuration, and so on. Although few systems can offer such features at present, most vendors will likely provide systems with such capabilities in a few years. If they do not, then they are bound to fail and may cease to survive because many large instrument suppliers faced a similar problem during the last decade and no longer exist as independent companies.

Although open system technology is well understood and widely applied in business and administration systems, the same cannot be said about control systems, for three reasons: (1) most C&I vendors do not offer such systems; (2) most C&I engineers are not conversant with systems theory, sampling theory, and information theory because these subjects are taught in special postgraduate courses only; and (3) most end-users resist changing, mainly because final decisions are made by mature mechanical or electrical engineers rather than control systems engineers.

This book explains, in simple terms and without applying complicated mathematics, the systems theory and its associated theories, together with the benefits of employing systems theory in the selection and design of C&I systems. I use examples from my own experience and projects in which I was responsible for the design and specification of the control system. My intention is to convince readers, especially those who are afraid of change, that the new C&I techniques will produce far better results than the traditional methods. It is utterly irresponsible to reject new solutions by using such arguments as the following:

- Nobody has used this method, and we do not wish to be the first.
- The existing method works satisfactorily, so why should we employ something different?
- This alternative is expensive, and there is no guarantee that it would produce long-term benefits.

Some of the present-day C&I systems are extremely powerful and flexible and, if applied properly, substantial improvements in productivity, operability, maintainability, and safety can be achieved. The doubtful engineer will not accept such possible benefits as adequate reasons for adapting new techniques.

This book has been divided into seven chapters, each describing an important aspect of C&I systems. Although the reader may choose a particular chapter(s) to study first, I do recommend that the first four chapters be read in the order they are presented. This approach will help familiarize you with the terminology and appreciation of the discussions.

Chapter 1 reviews the objectives of the C&I systems. In every project, various philosophy and study documents indicate the objectives and the means to achieve them. The possible benefits attributable to various objectives are discussed.

Chapter 2 presents an overview of instrumentation, control, and safety systems. The main components of the system (e.g., LAN, control unit, operator station, and host) are outlined. Interfaces to subsystems are discussed. System reliability and availability analysis are also covered. The latest developments in field instruments, subsea control, multiphase flow metering, and vessel control systems are studied.

Chapter 3 is devoted to the systems theory. Simple examples from various projects are used to explain the systems theory. Associated theories (i.e., control theory, sampling theory, and information theory) are also explained. Various system criteria, which are a direct result of the application of the systems theory, are explained. These include hierarchical system configuration, open system technology, universal operating system, minimum hardware variety, and software robustness.

Chapter 4 describes the integrated safety and automation systems (SAS). The development of true integrated systems would not have been possible without an understanding and application of the systems theory and the availability of powerful microelectronics. Early developments, the current status, and the future of integrated systems are discussed. The benefits of employing integrated safety and automation systems are also highlighted.

Chapter 5 explains the project engineering of C&I systems. Various topics such as conceptual studies, system requirements, vendor selection, process interface, operator interface, applicable software development, testing, training, commissioning, staffing, and documentation are included.

Chapter 6 provides some information about the application engineering of control systems. Typical control schemes, the traditional way of configuring them, and application of the latest control techniques are discussed. The use of fuzzy logic and neural networks to control such difficult processes as blowdown, paper moisture and caliber, varying deadtime, and sensitivity are explained. The application of simulation for control, testing, and training is also described.

Chapter 7 reviews six control systems from vendors of international repute. These systems were chosen because of my experience and involvement in evaluating them for North Sea projects. The exclusion of some systems does not mean that they are unsuitable for offshore industry or less capable than those included. Indeed, every system may be the most suitable for some particular applications. Because control systems hardware and software undergo frequent modifications, which may enhance their capabilities significantly, the reader is encouraged to refer to the vendor's documentation for the latest information.

A comprehensive list of abbreviations and terminology, references, and an index are provided at the end of the book.

Acknowledgment

Several of my colleagues have assisted me in preparing this book. I am especially grateful to Mike Tustin, Svein Hagen, and Sven Ove Abrahamsen for their valuable comments and guidance. I am also obliged to many of my present and past supervisors, who have helped me directly or indirectly with this book and my previous publications. They include Les Willcox (Manager of Control Systems Group, Brown and Root, Southern United Kingdom), Hans Vindenes (Manager of Control and Power Systems, Brown and Root, Stavanger), John F. Thompson (previous Manager of Instrument Group, Brown and Root, Southern United Kingdom), and the late Graham Standing (Manager of Applications Group, European Training Centre, Honeywell, Bracknell, United Kingdom). My sincere thanks to Shirley Peacock for typing, Liz Potter for help with the figures, and to my wife, Dorcas, for proofreading this book.

—Ghodrat Kalani

1

Objectives

1.1 INTRODUCTION

The aim of this section is to review the objectives we wish to achieve in the design of control and instrumentation (C&I) systems. These objectives must be defined before the start of the detail engineering. The objectives should be clearly stated and written in simple terminology so that all project staff, including junior engineers, can easily comprehend them.

After the objectives have been agreed on by all parties (i.e., contractor lead and systems engineers, client design, operation and maintenance engineers), the criteria and means of meeting the objectives must be specified. The documents that define the objectives and the ways they will be realized are various design philosophy reports. Design philosophies are discussed in the following section.

1.2 PHILOSOPHIES

Design philosophies for the C&I of a project are produced in an early stage of the conceptual study. The philosophy documents are the basis for the preparation of the C&I requirements specifications. For a large offshore project, the following typical philosophy documents may be necessary:

- Overall control and instrumentation philosophy
- Field instrumentation philosophy
- Process control system philosophy
- Emergency shutdown system/blowdown/process shutdown philosophy
- Fire and gas philosophy
- Metering philosophy
- Mechanical/electrical packages control philosophy

In an integrated safety and automation system, the following philosophies may be combined into one philosophy document:

- Process control system philosophy
- Emergency shutdown system/blowdown/process shutdown philosophy
- Fire and gas philosophy
- Mechanical/electrical control philosophy

It is also possible to include field instrumentation philosophy in the overall control and instrumentation philosophy. A philosophy document will normally consist of the following sections:

- *Introduction.* This section fulfills the following purposes: (1) the intent and a general description of the system are stated; (2) the objectives are explained; (3) general system requirements are highlighted; and (4) applicable standards and codes and a list of abbreviations and terminology are included.
- *General Philosophy.* The important features of the system are itemized. The stated overall features are the means and criteria to help achieve the objectives outlined in the introduction. This section should not extend beyond one page.
- *Overall Design.* Functional requirements (e.g., control, monitoring, management, and engineering) are described, then system topology is explained.
- *System Components.* Communication, operation, control, management, and auxiliary facilities are indicated.
- *System Interfaces.* The method of interfacing various subsystems with the main control system is included.
- *System Reliability and Availability.* A brief description of reliability and availability requirements and how they will be achieved is presented.
- *References.* A list of support documents, some figures, and diagrams is provided.

1.3 REQUIREMENTS

When operational research engineers are faced with an optimization project, they define the requirements in a mathematical model. The model includes an *objective function* and some *constraint functions*. The number of constraints depends on the size of the project and, for a large model, could exceed 1,000. The optimization model is formulated as follows:

$$\text{Maximize (or Minimize)} \quad a_1 x_1 + a_2 x_2 + a_3 x_3 + \cdots + a_n x_n$$

$$\text{Subject to:} \quad a_{11} x_1 + a_{12} x_2 + \cdots a_{1n} x_n > 0 \ (\text{or} < 0)$$

$$a_{21} x_1 + a_{22} x_2 + \cdots a_{2n} x_n > 0 \ (\text{or} < 0)$$

$$\vdots$$

$$a_{m1} x_1 + a_{m2} x_2 + \cdots a_{mn} x_n > 0 \ (\text{or} < 0)$$

This set of functions show a general optimization model; however, it cannot be easily applied to the design of a C&I system. Although some operational researchers may have tried, I have not heard of such a project. I will model a C&I system here to show readers the application of operational research to the design of control systems. This will also indicate the important requirements and may encourage some senior engineers/managers to use such a method in their projects. If readers wish to learn how to develop such models and solve them, Beale's *Mathematical Programming in Practice*[2] may be consulted.

The first step in building the model is to define the objective and constraint parameters. Typical parameters include the following:

x_1 = Cost of maintenance
x_2 = Cost of spare parts
x_3 = Cost of operation
x_4 = Cost of modifications
x_5 = Cost of losses (e.g., production, equipment, personnel)
x_6 = Cost of information technology
x_7 = Space and weight
x_8 = Flexibility
x_9 = Expandability
x_{10} = Operability
x_{11} = Cost of C&I

Readers should note that more parameters can be added or some parameters may be combined into one parameter. This fact becomes apparent during the building of the model and system analysis.

The second step in building the model is to define the objective function, which can be presented as follows:

$$\text{Minimize} \quad (x_1 + x_2 + x_3 + x_4 + x_5 + x_6 + x_7 - x_8 - x_9 - x_{10} + x_{11})$$

The coefficients $a_1, a_2, \ldots a_{11}$ have been dropped for the sake of simplicity; however, these can be included to emphasize the importance of some parameters. For example, if the cost of maintenance is critical (e.g., for remote satellite platforms), then the x_1 can be replaced by $2x_1$.

The third step in developing the model is to describe the constraint functions, as follows:

$$x_1 + x_2 + x_3 \leq C_1$$

This constraint implies that the total cost of maintenance, spare parts, and operation should not exceed C_1, otherwise the production will not be viable.

$$x_4 + x_{11} \leq C_2$$

This constraint states that the total cost of modifications and initial control and instrumentation system should not exceed C_2.

More constraints can easily be developed for the model. It is important that the objective and constraint functions are specified accurately. Otherwise, solving and analyzing the model may not produce meaningful results.

The last step is to analyze the model and find the optimum solution(s). When operational researchers tackle optimization problems, after finding the optimum solution(s), they normally carry out sensitivity analysis. Sensitivity analysis shows how sensitive the optimum solution(s) is to changes in one or more of the system parameters.

It seems doubtful if such models will be applied to C&I systems in the near future. The fundamental reason is that the development of accurate models is difficult and requires extensive research in order to comprehend the technical and financial parameters and their relationship. The reader may then question why I have introduced the subject if it may not be applied to the design of C&I systems at all. There are two reasons for discussing the optimization model here: (1) it is an interesting subject, and some mathematically minded C&I engineers may pursue it further; and (2) as stated earlier, it is to show the important requirements in the design of a C&I system.

The C&I system requirements are included in the objective function. As explained earlier, the number of requirements (parameters $x_1, x_2, \ldots x_n$) may be decreased or increased without affecting the overall system optimum solution. For instance, x_1 and x_2 (the costs of maintenance and spares) can be represented by one parameter. This reduces the degree of maneuverability in the analysis of the model.

Some of the stated requirements are essential for the survival of the project. For example, if the cost of maintenance and/or operation escalates, either the C&I system must be redesigned or the plant closed down. In this competitive economic and industrial world, profitability and return on the capital invested are of paramount importance. The engineers who are responsible for the design and selection of C&I systems should constantly remind themselves of this fact from the start of conceptual design to the end of detail engineering and commissioning. Those client engineers who can influence the process of C&I design should emphasize the importance of meeting the stated requirements and fully cooperate with the control systems engineer(s) to achieve a cost-effective design.

C&I systems hardware and software are changing constantly and rapidly by using the latest advances in electronics and information technology (IT). Vendors allocate a considerable amount of resources (labor, finance) to improve various features of their C&I systems. The driving force behind these changes is competition. To compete, one has to constantly look for methods of enhancing the system's capabilities and offer new and better solutions to control processes. The following aspects of C&I systems are improving continually:

- Standardization
- Speed of response
- Hardware variety
- Software portability
- User interface
- Memory
- Compactness
- Power requirement

- System integrity (reliability, availability, security)
- Control and logic algorithms

Unfortunately, most C&I engineers do not keep abreast of these changes and therefore cannot evaluate their impact on the project requirements. In order for an engineer to fully understand the project requirements and C&I system features and their relationship, he or she should have a thorough knowledge of the systems theory and relevant process experience. Although systems theory is taught in some BS and MS courses, it is the experience that is lacking and hard to gain.

Designing a control system is in some ways similar to driving a car. In order to drive safely, we must learn the theory (e.g., signs and codes and the operation of gears, brakes, steering, ignition, indicators, mirrors). Then comes practical experience; first with some lessons from an instructor, then with an experienced driver in a student driver car, and finally after passing the test and driving on ordinary roads for a few months, we can drive with confidence on highways.

The engineer who is responsible for the design of a major C&I system should have learned the systems theory in a degree course (i.e., BS, MS, Ph.D.), then trained under the guidance of an experienced systems engineer and designed a few small C&I systems. It is no more safe for a large C&I system to be designed by an engineer who is not equipped with systems theory and necessary experience than it is for a motorist without sufficient experience on sideroads to proceed up an interstate.

After the C&I system requirements are described in various philosophy documents, the specifications documents for all the project's facilities are prepared. The specifications will fully explain the criteria and details of ways and means of achieving the philosophy requirements. All the major interfaces (i.e., process, operator, engineer, maintenance, and electrical and mechanical packages) are included. The requirements should be clearly stated. The details of requirements should be structured such that when vendors respond with compliance statements, it is easy for the systems engineer to evaluate and choose the system(s) that match the stated requirements.

2

An Overview of Instrumentation, Control, and Safety Systems

2.1 INTRODUCTION

In this section, we will try to meet the following three objectives:

1. Adequately explain the main components of a modern integrated control and safety system.
2. Understand the important system interfaces.
3. Indicate the primary data requirements for the design of a control and instrumentation (C&I) system.

Integrated safety and automation systems (SAS) are much simpler than orthodox systems because of a reduced variety of hardware and software in the former systems. System interfacing is also easier and much more efficient in integrated systems than in nonintegrated systems. The reasons for simplicity and high efficiency will become clearer later on.

C&I systems are designed and built based on the data provided by the control systems engineer and the engineering team to the vendor. The accuracy, completeness, and unambiguity of the data are prerequisites for the successful design of a C&I system. Many C&I systems have been engineered poorly or have failed to meet important milestones because of a lack of adequate data.

2.2 PIPING AND INSTRUMENT DIAGRAMS

The piping and instrument diagrams (P&IDs) are by far the most important source of information for the design of a C&I system. This fact is recognized by most experienced process, control, instrumentation, systems, and project engineers. Shortcomings in P&IDs may result in a control and safety system with the following deficiencies:

- Errors that may never be discovered
- Errors that will be discovered because of shutdowns
- Underutilization of the system hardware and software
- Overengineering

Unfortunately, these deficiencies are not well understood by most engineers who are responsible for the design of P&IDs or are users of them. Perhaps the following two examples from the operation of an oil refinery (my own experience) and an offshore platform (experience by an application engineer) will help clarify the situation.

During my work at Shiraz Refinery (in Persia) a major shutdown was caused by a fire at the desalination vessel. The desalination vessel was fed heated crude oil, and it separated gas (from the top), oil (from the middle) and water (from the bottom) for further processing. An increase in the level of water (water–oil interface) together with a decrease in the level of oil (oil–gas interface) resulted in heated water (130°C) penetrating into the gas areas. This caused a large and sudden increase in pressure and subsequent release of gas, oil, and steam into the atmosphere. The release of hot oil and gas and contact with hot equipment surfaces created a major fire, which damaged several pieces of equipment and caused a long total refinery shutdown.

An application engineer working in Britoil North Sea offshore platforms told us of his experience with the plant's control during a training course on TDC 2000. He paid regular visits to the platforms, and in every visit he would find several control loops in manual mode. Every time he noticed this problem, he would switch the control loops to automatic mode and instruct the operators that the loops should remain in automatic mode.

These examples demonstrate that the C&I of the refinery desalination vessel and offshore platforms had shortcomings. P&IDs are the primary design document for the configuration of a plant C&I. By reviewing the P&IDs, I discovered that the control and shutdown system of the desalination vessel was poorly designed. The shutdown system should have been a high-integrity pressure protection system (HIPPS) configuration. It should have been equipped with preshutdown warnings in order to facilitate manual intervention before the release of gas and steam into the atmosphere.

In the case of the offshore platforms, I did not have the chance to study the P&IDs. The application engineer believed that the operators switched the control loops to manual mode for no good reason; however, I have no doubt that either the control loops configuration was poorly designed or the proportional integral derivative control (PID) parameters were set incorrectly. The former situation is the more likely cause of the problem. Plant operators are normally highly knowledgeable and responsible people, who would not place control loops in manual mode unless the controllers created instability in their normal mode.

P&IDs also provide the necessary information to develop an input/output (I/O) database, graphic displays, mimics, alarm displays, reports, trends, and so on. It is obvious that erroneous P&IDs or their late completion will cause short-term and long-term problems in the C&I system's design. Various documents (e.g., instrument index, data sheets, hookups, and loop diagrams) require information from P&IDs.

P&IDs can be divided into two broad categories: main process P&IDs and package P&IDs. Although the latter is more problematic than the former, both types require special attention. Unfortunately, inexperienced or unqualified instrument engineers are often assigned to the supervision and preparation of P&IDs. It is vital that a senior instrument or control systems engineer should be assigned to help with the design of P&IDs. The preparation of package (e.g., generators, compressors) P&IDs is normally delayed for the following reasons:

- Late order placement for the package
- Lack of cooperation by package vendors
- Assignment of inexperienced and/or unqualified instrument engineers to packages

Normally, the last item is the root-cause of the problem. This lack of experience can lead to late preparation of documentation and testing. Because the packages interface with the main control system via serial links and require the necessary database and graphic displays, the poor package P&IDs can cause serious problems in the configuration and testing of the main control system.

2.3 SAFETY AND AUTOMATION SYSTEMS

2.3.1 System Configuration

Figure 2–1 shows the main components of an SAS. This is a simplified system block diagram. The block diagram for a typical offshore complex will require several pages of A3 size and will indicate all subsystems (i.e., electrical, mechanical, telecoms) and their interfaces and all the items of equipment, including auxiliary facilities such as printers, keyboards, and monitors. It is recommended to use one page for each major area (e.g., satellite platform, a subsea template, power generation module, utilities, compressors set).

The block diagram should be easily comprehensible and clearly show system hierarchy and all important interfaces. The control system block diagram is a fundamental document, which is constantly reviewed and updated from the moment it is conceived. The main constituents of the SAS (i.e., local area network [LAN], operator station [OS], control unit [CU], and host computer) are briefly described in the next four sections.

2.3.2 Local Area Network

The LAN, bus, or data highway, as it is called by some control system suppliers, has undergone substantial changes since the mid-1970s, when the TDC 2000 was introduced. At those early times, not only were the data highways very slow, but each vendor developed its own proprietary protocol, which made it virtually impossible for a third-party system to interface the highway adequately.

Early distributed control system (DCS) highways had a speed of 100–250 kBits/second, while modern systems employ 5–10 MBits/second LANs. Various limitations such as the speed of transmission, length of highway cables,

10 | INDUSTRIAL PROCESS CONTROL

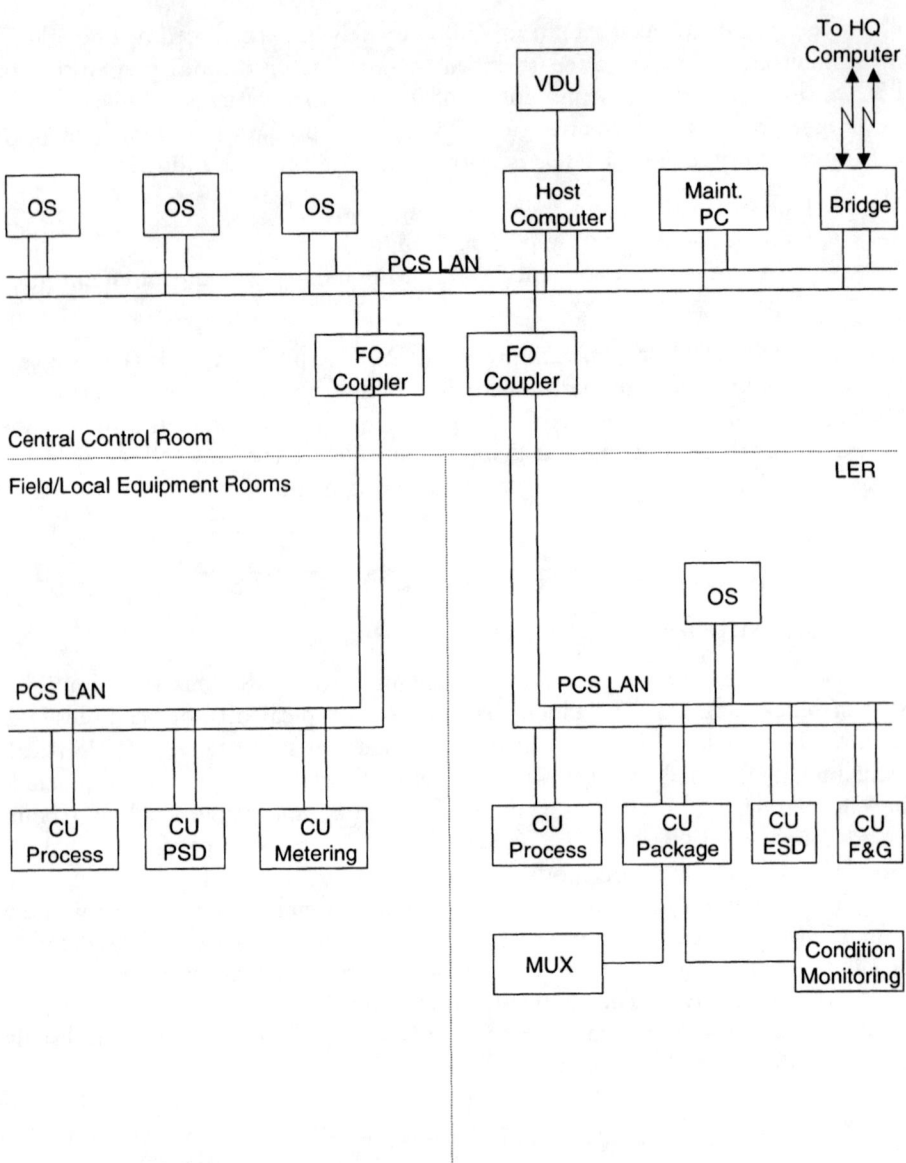

Figure 2–1   Control and safety system block diagram.

number of branches, and the number of nodes yielded control systems with limited rangeability, flexibility, and expandability. The present-day control system LANs have the following features:

- Conform to industry standards (i.e., IEEE 802)
- 5–10 MBaud transmission rate
- Branching or bridging that offers a highly structured and robust hierarchy
- Extension via fiber-optic (for harsh environments) and telemetry (for remote areas)
- Virtually unlimited length
- A large number of nodes
- A large number of I/Os; some systems can handle 100,000 I/Os

The LANs offered by control system vendors conform to one of the three IEEE standards, namely Ethernet (IEEE 802.3), Token Bus (IEEE 802.4), or Token Ring (IEEE 802.5). Few systems employ Token Ring because of the protocol's complexity and limited advantages.

I prefer Ethernet because it is easier to implement and far more popular in the computer industry than the other two standards. Nearly all computers, including PCs, can offer Ethernet, whereas that is not the case for Token Bus and Token Ring. There is no technical advantage between using Ethernet or Token Bus, although the suppliers of these protocols claim that their standard is better than the other. The only advantage of Ethernet over Token Bus and Token Ring is that the former yields a more open system environment.

For a recent project, in which I was responsible for the design of the SAS, the chosen system provided Token Bus. The system needed to interface with more than 30 major packages, which employed their own programmable logic controllers (PLCs) and computers. Most packages could offer Ethernet, whereas none of them could provide Token Bus. Consequently, the process control system (PCS) had to interface with various packages, including the emergency shutdown system (ESD) and fire and gas system (F&G), via RS 485 serial links. Approximately 100 serial links were designed, tested, and commissioned for the project. Figure 2–2 shows package control system interface to the main control system (PCS) for Ethernet and Token Bus LANs.

The serial links and associated gateways, if not designed and planned properly, will increase the cost of the system significantly and may hinder the commissioning and startup of the project. Gateways require space, which is a rare commodity in offshore platforms, and reduce the reliability of the system. In Figure 2–2, the system (a) is superior to system (b), not just because the former does not require a gateway and a CU to interface the LAN. There are many reasons for the superiority of system (a), which may not be obvious to the inexperienced engineer. In addition to savings in cost and space, and major advantages in reliability offered by system (a) versus system (b), the following subjects should also be considered:

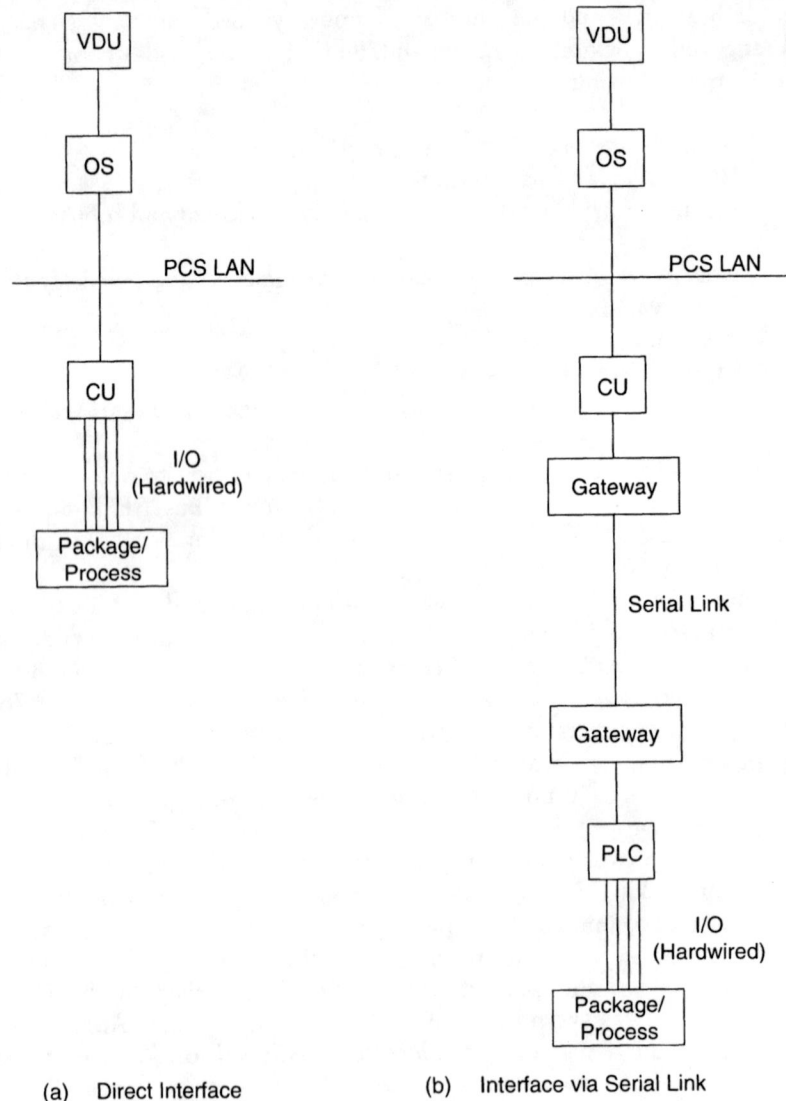

(a) Direct Interface (b) Interface via Serial Link

Figure 2-2 Package PLC–PCS interface.

- *Engineering effort.* The development of an interface between package PLCs and gateways (e.g., RS 485) could prove problematic. Not only do we require software and application programming for the interface, but we may also need new or revised hardware, whose acceptability will be questionable. I have witnessed the testing of many serial links, and they are rarely error-free. They often need modifications and retesting. Even after successful factory acceptance testing, they may need modifications during commissioning.

- *Speed of response.* In the system (a), the process I/O can be exchanged with a LAN (or operator station) within 1–2 seconds; however, in the systems using gateways, the delay could be as high as 20 seconds. Unfortunately, the slow response problem does not reveal itself until testing, which may be too late or difficult to rectify. In some cases, the engineer may not appreciate the severity of the problem and may ignore the shortcomings altogether.
- *Loss of information.* Because of a low speed of response, vital information may be lost. For example, in a serial link to an antisurge controller or a steam power generator PLC with a response time of 2 seconds, pressure, flow, or level (steam–water interface) changes may easily be lost. The process parameters can change several times within 2 seconds, especially during process disturbances. This subject is further discussed under sampling theory.
- *Lack of expertise.* The serial interface is usually a small portion of the total system in mechanical/electrical packages. Consequently, the package vendor either assigns one software engineer for all protocol software development, or subcontracts the task to a third party. Both methods can and often do prove problematic. In the former case, because software engineers are normally young, it is possible that the assigned person will leave the firm during the course of the project. A new engineer needs to be trained and may take a long time to become familiar with the system. The second method (i.e., subcontracting the interface software to another company) creates two problems: (1) lack of adequate control on the software company, and (2) the fact that software houses are normally small firms and may not have adequate expertise in the interface protocol.
- *Hardware incompatibility.* Because the protocol implementation may require firmware and hardware modifications to the system hardware, enhancements to the vendor hardware may create difficulties during system expansions and upgradings. For instance, if the vendor upgrades the PLC processor, and the end-user wishes to replace the old processor (to enhance the system performance), or needs another PLC for plant expansion, the PLC may not be compatible with the existing gateway. If the control systems engineer is not aware of this incompatibility, delays in the plant startup could be substantial.

The problems associated with hardware incompatibility and lack of expertise can be best illustrated with difficulties we faced during an interface test where I was the control systems engineer in a North Sea project. For the second phase of the project, a compressor set, to boost the reservoir pressure by gas injection, was ordered from the same vendor as of the existing compressors. Although the compressor supplier was a large organization, it had a small subsidiary, which designed and developed control systems for compressors.

Because the PCS and compressors control systems for Phase 1 and Phase 2 were identical (the same vendors and the same hardware/software), I assumed that the serial interface test would go smoothly. The contractor instrument engineer and I (as the client control systems engineer) went to the United States, where the compressors control system was being built, to witness the serial interface testing. We arrived at the factory on Monday morning to start witnessing the test; however, the vendor's representatives said that they had not completed their in-house test, and the factory acceptance test (FAT) would start on Tuesday morning. We reported on Tuesday morning for the FAT, but we were told that there was a minor software problem. The vendor's representatives were confident that they could rectify the software error and be ready for the FAT on Wednesday morning. The same story was repeated on Thursday and Friday.

On Friday morning I asked for a meeting with the vendor's software and hardware engineers to discuss the problem and decide on the next course of action. We were told that the hardware and software (application and firmware) were similar to those of Phase 1's control system. The engineers said that they had replaced every card in the system, but to no avail. They were planning to work over the weekend to have the system ready for test on Monday morning.

I did not accept the vendor's proposed course of action on the basis that because they could not resolve the problem in the previous five (or more) days, there was no guarantee that they would resolve it if they repeated the same checks. Then the vendor's people asked my advice on what steps they should take. I asked the software engineer if he had written the serial interface software for the control system of Phase 1. He said he had joined the company recently, and the engineer who had originally developed the interface software had left the firm a few months previously. I suggested that they should ask the former employee for help by attending tests over the weekend. The vendor was reluctant to seek help from an ex-employee because this meant admitting defeat; however, they did contact the engineer and explained the problem to him and asked for assistance. He discovered that some of the PLC electronic cards were of different revisions from those he used when he developed the serial interface protocol. Some cards had been recently revised and consequently were not compatible with the others. The new cards lost their compatibility because the interface software, which was written for the old cards, was not suitable for the new ones.

This example shows how serial interfaces can delay the progress of the project and how a minor software/hardware incompatibility can cause frustration and consume engineers' valuable time. The application of serial links in PCS will result in some or all of the following shortcomings:

- Degraded response performance
- Reduced reliability
- Reduced monitoring capability
- Reduced flexibility
- Increased testing effort

- Increased costs
- Delayed equipment delivery

2.3.3 Control Unit

The old-generation DCS offered several control and data acquisition units. Each unit provided for different aspects of control, monitoring, and safety requirements. For instance, the TDC 2000 control and data acquisition units were as follows:

- Basic controller (BC)
- Extended controller (EC)
- Multifunction controller (MFC)
- High-level plant interface unit (PIU)
- Low-level PIU
- Low-energy PIU
- Tricon or a similar PLC (for safety systems)

The fundamental problem with such a large number of components is that they require different hardware, software (database configuration), training, and expertise (maintenance, engineering). The new generation of control systems provide only one CU. The CU is used for control, data acquisition, and logic (ESD, F&G, sequence). A CU should offer the following capabilities as standard:

- Continuous control
- Batch control
- Logic
- Advanced control
- Simulation
- Neural network and knowledge-based systems
- A PCS-oriented programming language
- Dual or triple redundancy
- High scan rates (1–10 msec)
- High-resolution time stamping (1 msec)
- A communication processor that can handle popular protocols

The application software and database configuration tools should be engineered such that process and instrument engineers can easily program the system without the need for substantial training. In other words, it should be user-friendly. The process interface of the CU is critical, and a well-designed CU will support the following I/O types:

- Analog input; 4–20 mA, T/C, RTD, 12-bit or better A/D conversion, 32-bit storage
- Analog output; 4–20 mA
- Digital input; status, alarm, sequence of events (1 msec resolution), normally open or closed, volt-free, powered (24 VDC, 110 VAC, 240 VAC, 50 Hz, 60 Hz)

- Digital outputs; volt-free and powered
- Pulse input; configurable frequency to 16,000 Hz
- Pulse output (stepper); configurable to 1,200 steps
- Fieldbus

The CU alarms on various parameters both standard and programmed should be provided. Typical alarms are as follows:

- Process Variable (PV): H, L, HH, LL, ROC, Bad PV
- Deviation: H, L
- Digital input: open, closed
- Digital output: command disagree

2.3.4 Operator Station

The OS is the primary interface for operators, engineers, and management. It provides all of the necessary graphic, mimic, tabular, trend, reports, configuration, and diagnostic displays to satisfy all the system users' requirements. The operation and monitoring displays are based on a strong hierarchy in order to provide efficient operator interfacing. The operator should be provided with adequate information at all process conditions. These conditions include normal, disturbances, partial shutdown, total shutdown, startup, and maintenance. Bombarding the operator with too much information is as undesirable as presenting him or her with too little data.

The following displays are available in most control systems:

- Overview
- Area
- Group
- Detail
- Trend
- Configuration
- Diagnostics
- Alarm summary
- System (LAN) status
- Scratch pad

Overview displays can be graphic or semigraphic, showing a major part of the process (e.g., a satellite platform or a boiler plant) or the total plant. Overviews are the primary displays during normal operation. Any process disturbance will be indicated early enough for operators to take remedial action in order to prevent plant shutdown.

Overview displays work on the principle of "report by exception." This allows an overview display to be linked to hundreds of process parameters. At the early stages of process disturbances, one process variable (e.g., a vessel level or pipeline pressure) will indicate a deviation from the setpoint. This may be indicated by the movement of a vertical bargraph below or above a horizontal line (line of zero deviation) or change of color of a vessel or a pump. The operator will be prompted to the imminent alarm conditions and to the appropriate area/group display for detailed information in order to take remedial action.

If the operator fails to restore the process to normal conditions quickly, several parameters may exceed their warning or alarm limits. In such cases, the operator may be prompted to several group/area/alarm displays, so consequently the overview display will lose its significance. Software packages to filter alarms and help operators in situations where many alarms and warnings occur simultaneously (i.e., alarm floods) are available. The packages employ neural networks, expert systems, or similar techniques. Normally, a shell will be procured that will need to be trained on a continuous basis. Operator actions in known situations are fed to the shell in order to educate it. Historical data are valuable for the package training.

Area and group displays provide detailed process information and the means to operate process plant equipment (e.g., open a valve, shut down a pump) by the use of keyboard, trackerball, touchscreen, mouse, and so on. Moving from group/area displays to overview (or vice versa) should be easy, and normally by one or two keystrokes. The operator station should provide standard overview, area and group displays, and a means of generating project-specific displays to satisfy operator requirements.

Detail, configuration, and diagnostic displays are standard OS displays. They present detailed data required for such tasks as tuning, calibration, database development, and maintenance. Trend and average displays provide historical data. Alarm summary indicates active, acknowledged, and disabled alarms in a chronological order. Scratch pads are used by operators/engineers to enter offline information such as notes, messages, reminders, and the like.

The present-day operator stations are normally UNIX and Windows-based systems, which will satisfy any project requirements. Some control systems offer a powerful OS with capabilities of supporting more than 100,000 process I/Os and refresh time of 1 second. Windowing, zooming, rolling, and overlaying are standard features of these systems, which should be fully utilized in order to achieve high efficiency in plant operation and monitoring.

Handling of alarms and warnings is one of the most critical aspects of the operator interface. During process upsets, alarm floods may create an unmanageable situation if the control system alarm configuration is poor. Long shutdowns, production loss, damage to process equipment, or major hazards may occur as a result of mishandling of alarms. Such techniques as prioritization, filtering (masking), and hierarchical design should be employed in the design of alarm systems. The use of neural networks, fuzzy logic, and expert systems in the design of alarm systems in the future will yield substantial benefits. Normally, such software packages can be added to the control system after commissioning, provided the system is an open one.

2.3.5 Host Computer

Host computer is the highest level of hierarchy in the control system, although some companies may employ a mainframe or super/mini computer in their headquarters for the management of the plant(s). This is especially the case when several process plants are managed from the company headquarters. Radio links, satellite, fiber-optic, or telephone lines may be used to connect the host computer(s) to the headquaraters mainframe.

The host's primary function is to provide a global database. The host is equipped with a large mass memory (e.g., several GBytes of hard disk or tape). Process parameters, including averages and totals for several years, may be archived by the host. Hosts are normally UNIX-based, which makes it possible to apply most software packages written by various vendors. Such tasks as optimization, simulation, preventive/predictive maintenance, statistical analysis, stock control, and management reports are within the domain of the host.

2.4 INTERFACE TO FOREIGN SYSTEMS

2.4.1 General

Since the introduction of microprocessor-based distributed control systems, the interface to foreign systems has posed ever-increasing difficulties to control and instrument engineers. In a recent offshore project, in which I was responsible for the design of the control and safety systems, the control system had more than 70 serial links for interfacing with other systems and more than 20 internal serial links (between PCS, ESD, and F&G systems). Most of my effort, in addition to some of my engineers' as well, was expended on tackling serial interface problems.

The original control system specification called for an open system, where the PCS control unit would be used to control all packages; however, the client changed course, and hence numerous types of PLCs with the accompanying serial links were introduced. Employing so many PLCs and their serial links to the plant PCS created the following problems:

- Substantial demand on the project's control systems engineer
- Substantial demand on the PCS vendor's systems engineer
- Poor engineering of mechanical/electrical packages control systems because most instrument/electrical engineers assigned to various packages had no experience or understanding of PLCs or their serial interfaces
- Slow speed of response in many serial links
- Substantial cost of testing and retesting serial interfaces
- Increase in the variety of hardware and software
- Poor reliability of serial interfaces because a gateway (sometimes two or three in series) is normally required to provide the protocol conversion between PCS and PLCs

- Delay in the delivery of equipment. Alternatively, modifications and tests are carried out at site or offshore at extra cost.
- Some smaller companies do not employ experienced software engineers to develop the serial interface protocol.

As explained in Section 3.1, interfacing is the most important constituent of the systems theory. In order to fully satisfy the requirements of the systems theory, no serial links should be used to interface foreign systems to the plant PCS. Two options are then available: either to (1) use a PCS control unit for the control of the packages, or (2) employ a PLC that can interface the PCS LAN directly (i.e., without a gateway). The former option is preferred.

When serial links cannot be avoided, the following guidelines should be observed:

1. Foreign PLCs should use the PCS communication protocol, although an industry standard (e.g., MODBUS) may be preferred.
2. Protocol characteristics, speed, parity, and hardware should be standardized within the project.
3. If the chosen protocol offers various options, the best options should be selected. For example, if the protocol offer RS 232 and RS 485, the latter should be specified. If 19.2, 9.6, and 4.8 kBaud are available, 19.2 kBaud is the preferred rate.
4. A test of serial links should be carefully designed and planned.
5. A serial links schedule to indicate necessary information (e.g., the number of I/Os, nodes, multidrops, redundancies, and dates of tests) should be provided and updated frequently.
6. A senior systems engineer, preferably a member of the PCS team, should take responsibility for serial interfaces. This engineer should have control and supervision over all instrument engineers who are responsible for packages with serial interface to the PCS.

I have been responsible for the design, detail engineering, and testing of more than 100 serial links since the mid-1980s. Rarely did any of the serial links work error-free when we first tested them. Many needed revisions to hardware, software, firmware, and database, and had to be retested; however, all of them worked satisfactorily in the end. Some serial links, engineered by some of my colleagues, failed to meet the project requirements (e.g., too slow, crashed frequently, could not handle all I/O) and consequently were replaced by parallel hardwiring at the last minute. The problems associated with serial links are horrendous and usually not appreciated by most C&I engineers, especially older ones. My recommendation is not to use serial links in the first place, but if you have to, design and plan them carefully and assign a young senior and experienced control systems engineer to the task, and provide him or her with full support.

2.4.2 Object Linking and Embedding (OLE) for Process Control

In a large process plant, several major packages may need dedicated control systems provided by their vendors. In an oil and gas production plant, typically generators, compressors, switchboards, heating, ventilating, and air conditioning (HVAC), and metering (fiscal, allocation, multiphase) are equipped with dedicated control systems. Additionally, where the plant includes subsea well control pods, vessel control (e.g., for a floating production, storage, and offloading vessel [FPSO]) and cargo/offloading systems, they will also be provided with dedicated control systems.

Traditionally, all of the aforementioned control systems would interface with the PCS via serial links (MODBUS RS 485). Serial links have serious drawbacks (e.g., very low speed, low capacity for data exchange, short cable length, and considerable troubleshooting requirements). With the introduction of integrated SAS, the requirements for serial links have been reduced substantially. The main objective of integrated SAS is to minimize the hardware variety. By using the plant SAS control unit for the packages (a dedicated SAS control unit will replace the vendor-provided unit control panel), not only does the spare parts management and maintenance become much more efficient, but there will also be no need for troublesome serial links.

The introduction of OLE for process control (OPC) has also simplified the control system interfacing. Package control units and management computers (e.g., office PCs, simulation servers, store/maintenance servers) can interface with SAS via the OPC net, instead of serial links. OPC provides a standard method for data exchange among various applications. OPC uses Ethernet, which is the industry standard and is accepted by all SAS vendors. OPC networks have several major advantages over serial links. Serial link speed is normally limited to 19.2 kbps, whereas OPC nets, using Ethernet, have a standard 10 mbps speed (100 mbps is also available). Ethernet also allows much larger cable lengths and different types of media (e.g., twisted pair, coaxial, fiber-optic).

OPC, which is built on Microsoft's OLE/COM technology (object linking and embedding/component object model), uses a client/server configuration. The OPC server collects data from the SAS control units, including package control units (unit control panel [UCP], PLC, remote terminal unit [RTU], etc.). The client (operator stations, historian, PCs) requests data exchange from the server, when required. In some modern control and safety systems, the client/server configuration may not be visible. For example, a package control unit may have its OPC communication cards directly connected to the SAS LAN, and the OS requesting data exchange directly (i.e., no OPC server). This configuration, which is used by modern control and safety systems, is more reliable than the original client/server configuration because it does not rely on an OPC server. Figures 2–3 and 2–4 show the two types of SAS–OPC interface configurations.

The OPC data access is flexible (unlike serial links), and the following types are available:

- Synchronous data read
- Synchronous data write
- Asynchronous data read
- Asynchronous data write

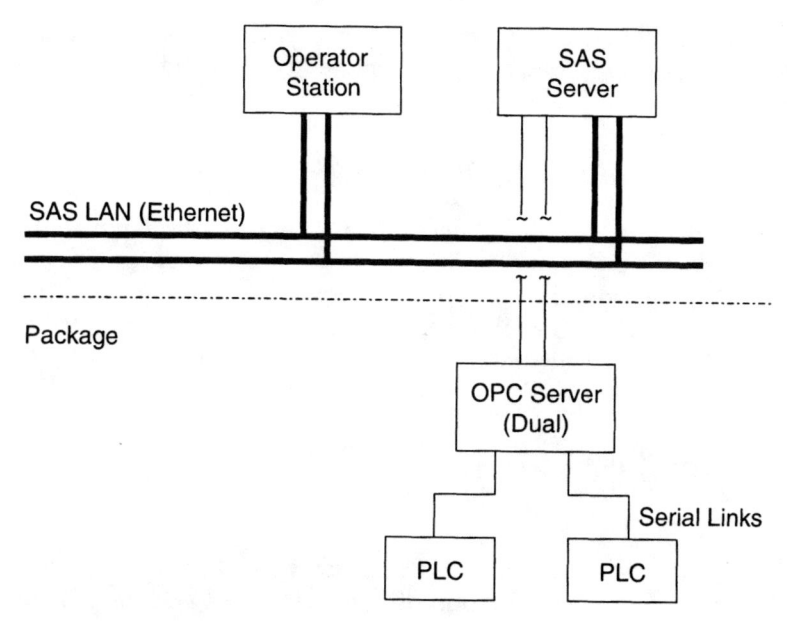

Figure 2-3 OPC interface using servers.

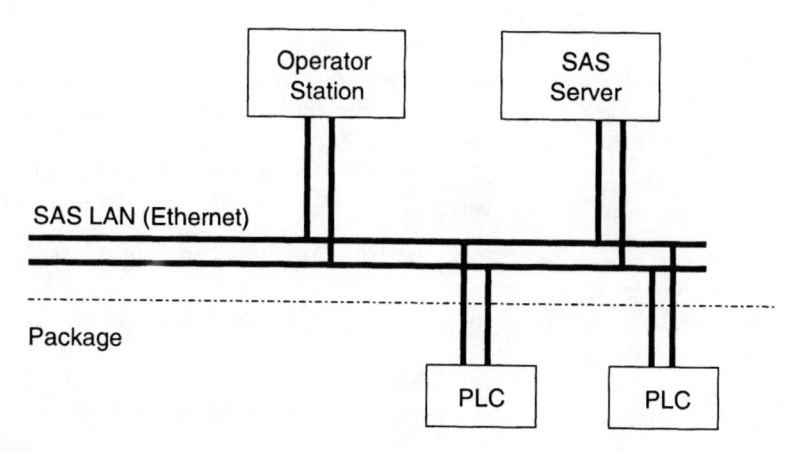

Figure 2-4 OPC interface direct.

- Process data change notice (if data are changed, memory is updated and data are sent to clients)
- Process data refresh (if data are changed, reads data and sends it to clients)

Is OPC the panacea? The primary advantage of OPC is the replacement of slow, troublesome serial links with a much faster interface. For the orthodox control and safety systems, this is obviously a major improvement; however, in integrated control and safety systems, there is no need for serial links and, consequently, little need for OPC. In an integrated control and safety system, the SAS control unit is employed for all applications (e.g., PCS, ESD, F&G, compressor and generator packages control, metering, subsea). In this case, the only application for OPC may be the SAS–PC interface.

In support of the OPC data exchange standard for Ethernet (OPC DX), the Fieldbus Foundation, Profibus International, Control Net International, and the Open Device Net Vendor Association have agreed to support a new OPC standard. The object of this new task is to allow data exchange between control units from different vendors via Ethernet transmission control protocol/Internet protocol (TCP/IP). Participating companies are Emerson, Rockwell, and Siemens.

2.5 FIELD INSTRUMENTS

2.5.1 General

The use of microelectronics in field instruments (transmitters, valves) has opened a new era in the application of interfacing C&I systems. The 4–20 mA twisted pair used in the previous generation of analog instruments has two major restrictions: (1) each field instrument requires a dedicated pair of wires to the control room; and (2) only one piece of information per pair of wires is possible (i.e., process variable or controller output).

Smart field instruments employing profibus or fieldbus do not impose these restrictions.[3-5] Many instruments can be connected to the profibus or fieldbus, which can facilitate data exchange between field instruments and the SAS in the control room, or between the field instruments themselves. The combination of profibus or fieldbus and smart instruments yields the following advantages:

- Reduced cabling, cable trays, marshalling, and junction boxes
- Reduced space requirements
- More robust transmission by the use of digital data exchange and fiber-optic cables
- Better flexibility and expandability because nodes can be easily added to the existing fieldbus or profibus (whereas adding analog instruments could be difficult because of the requirement for extra cables, cable trays, marshalling cabinets, etc.)
- Substantially more information than equivalent analog instruments (e.g., diagnostics messages, performance data, calibration and tuning [reconfiguration] from SAS)
- Local data exchanges (e.g., local smart transmitters and smart valves can provide local PID control or logic control [interlocking])
- Digitization and characterization (preprocessing) of process parameters reduce the SAS control units' processing burden substantially
- PCS control units can handle far more I/O process because of reduction in marshalling and CPU processing requirements

- Reduction in engineering, drafting, documentation, and CAD requirements
- Reduction in overall C&I costs

In the following two sections, the profibus and foundation fieldbus (FFBus) will be briefly discussed.

2.5.2 Profibus

Profibus is based on the European standard EN 50170. Its physical properties, access methods, and user protocol conform to International Standards Organization (ISO) standard layers 1, 2, and 7, respectively. Its applications include field instruments (i.e., replacing the 4–20 mA medium) and small cell networks (i.e., instead of a local PCS bus or a serial link). Profibus was developed with the cooperation of several companies, including ABB, AEG, Honeywell, Siemens, and some universities. It can be used at both the field instruments level and the control system level. Figure 2–5 shows a profibus network, employing SIMATIC control units.

The extent of profibus application will be vendor-dependent. Although profibus offers many advantages over the 4–20 mA system, it has not become as widely popular as one would expect. It has not found the same degree of acceptability among vendors and users as FFBus has. Various transmission media are possible via profibus, including the following:

1. *Twisted pair cable*, especially designed cables, are available as follows:
   - Standard cable for ordinary applications
   - Robust cable for use where chemical hazards or mechanical stress may exist
   - Flame-retardant and halogen-free
   - Flexible cable for use in moving equipment
   - Fast-connect special cable for food industry
   - Fast-connect special cable for laying underground
2. *Fiber-optic cables* with glass fiber or plastic fiber. Plastic fiber-optic cables are primarily for indoor use. They can be up to 80 m long, although special cables are available for longer lengths. Glass fiber-optic cables for outdoor or forced movement applications are available. Fiber-optic cables for long distances (80 km or longer) are possible, although repeaters may be required. The advantages of fiber-optic cables are as follows:
   - No interference by electromagnetic noise
   - Can be used for very long distances (more than 50 km)
   - Provides galvanic isolation
   - High transmission rate (up to 12 MBaud is possible)
3. *Wireless transmission*. Infrared modules are used to provide interface between profibus segments. Speeds of up to 1.5 MBaud and distances up to 15 m are possible.

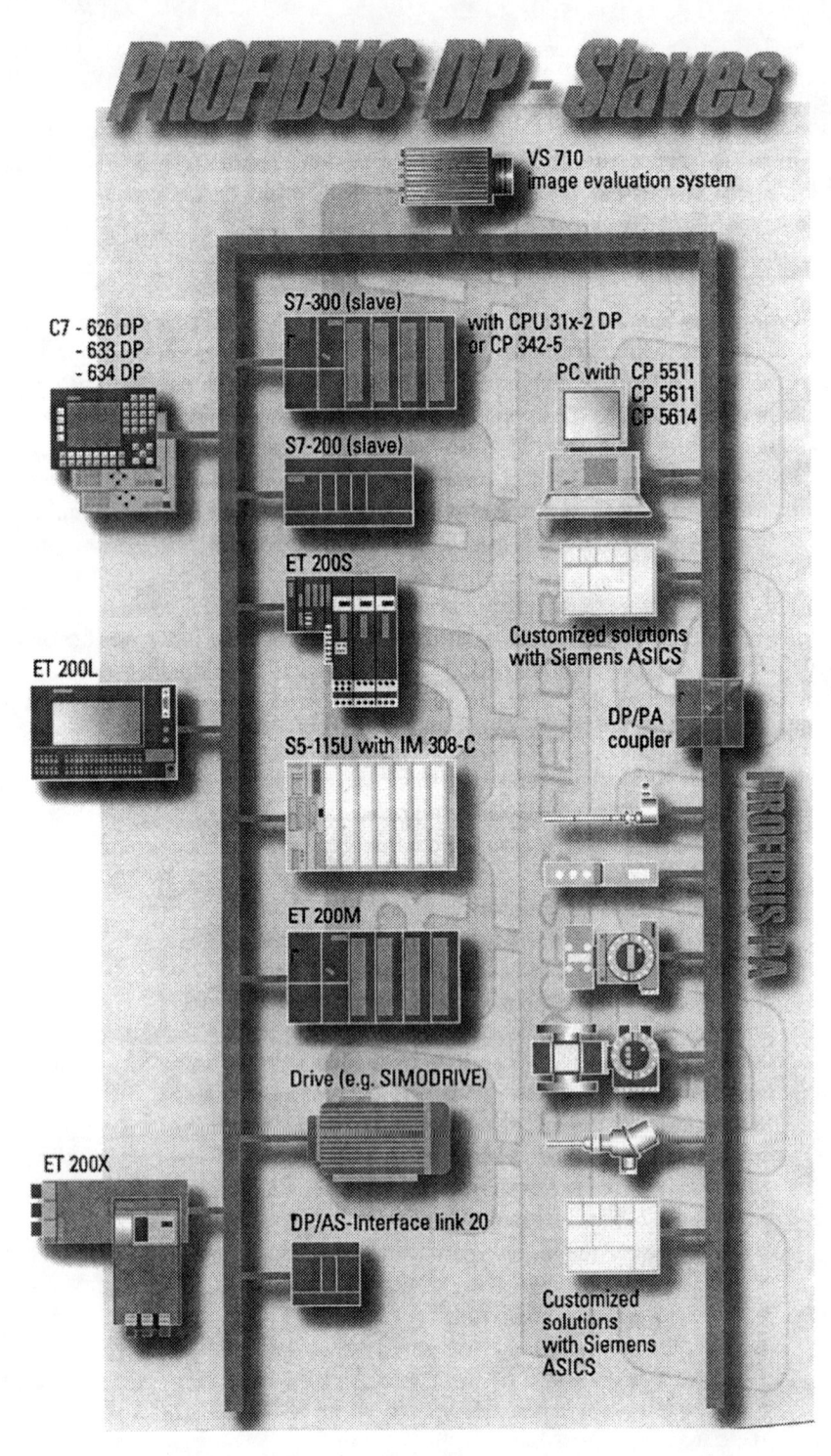

Figure 2–5   Profibus network (courtesy of Siemens).

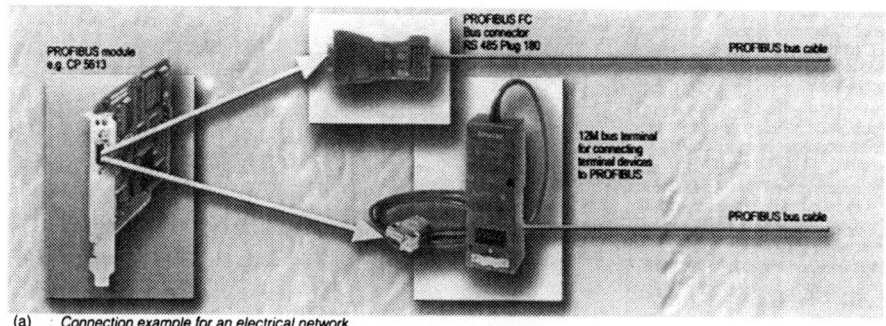

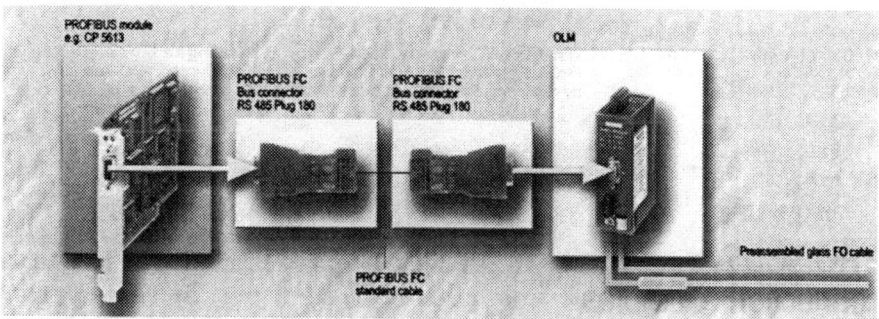

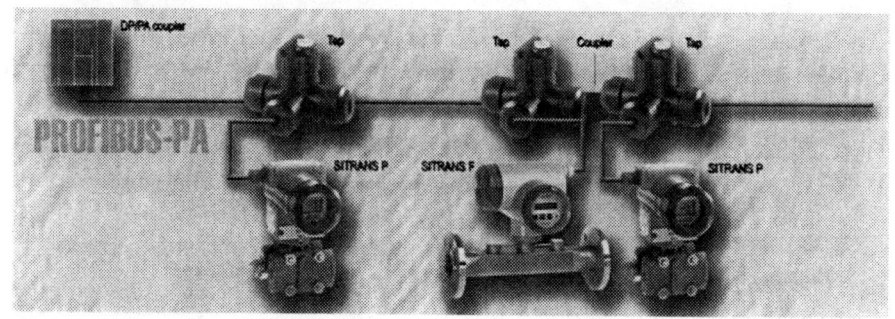

An Overview of Instrumentation, Control, and Safety Systems | 25

Figure 2–6 Profibus connecting elements (courtesy of Siemens).

Figure 2–6 shows miscellaneous profibus elements for interfacing various parts of the system. Instruments (transmitters, valves) by other manufacturers can be connected to profibus, which may be monitored/controlled by a PCS provided by another vendor. The main components of the profibus are the following:

- Repeater
- Optical bus terminal (OBT)
- Optical link module (OLM)
- Interface link module (ILM)
- Profibus–profibus link

- Profibus connectors (straight, 90-degree, fast and split, split coupler, split tap)
- RS 485 connector
- Fast-connect special cable for laying underground

2.5.3 Foundation Fieldbus

At present, several fieldbus protocol standards are available, which have been used in several projects (process plants, large buildings). Examples of these protocols are as follows:

BITBUS	IEEE 1118
CAN	ISO 11519/11898
CEBUS	EIA-IS-60
IEC Fieldbus	IEC 1158-2
Interbus	DIN E 19258
Profibus	DIN 19245 T1 to T4 / EN 50170/ IEC 1158-2
P-NET	EN 50170
World FIP	EN 50170 / IEC 1158-2 / NF C46-602 20 07

Foundation fieldbus (FFBus) is a digital communication system, whereas some of the previously listed protocols are hybrid systems. In hybrid systems, digital data are superimposed on the conventional 4–20 mA signal. The communication system standard for hybrid systems is normally vendor-dependent. In the FFBus systems, not only are these restrictions removed, but other improvements are added as well. A summary of improvements, which FFBus will offer as it becomes more widely applied, are given as follows:

- Higher communication speed (e.g., 10 MBaud or even 100 MBaud)
- Higher number of nodes per branch to reduce the cabling and termination effort
- More efficient communication (e.g., the use of report by exception between nodes)
- More intelligent field instruments. This will allow the transfer of more control and logic functions to field instruments and use control units for more complex tasks (e.g., multivariable control, supervisory control, optimization, simulation).
- Better diagnostics and predictive maintenance in field instruments
- More reliable control system because of better maintenance and higher distribution control
- Faster control system response (to avoid aliasing) because local instruments sample process parameters, rather than remote control units. This also relieves control units from the chore of sampling process signals
- Saving in hardware (e.g., cabling, junction boxes, cable trays, termination glands, I/O cards, IS barriers, and cabinets). Savings of 15 to 30 percent are possible

- Higher accuracy because process parameters are sampled locally and transmitted digitally to local/remote control units
- Major improvements in system commissioning; savings in time and costs
- Reduction in documentation (the number of loop diagrams, hookups, termination schedules, etc.)
- Possibility of using multifunction instruments, where one transmitter measures multiple variables (e.g., a coriolis meter can measure flow, density, and temperature)
- A high degree of interoperability among system computers and instruments from different vendors

The standardization of FFBus has been provided by the following organizations:

- ASHRAE (American Society of Heating, Refrigeration, and Air Conditioning)
- ANSI (American National Standard Institute)
- BSI (British Standards Institution)
- CENELEC (Committee European de Normalisation Electrotechnique)
- DIN (Deutchers Institut fur Normung)
- EIA (Electronic Industries Association)
- IEC (Institute of Electrotechnical Commission)
- IEEE (Institute of Electrical and Electronic Engineers)
- ISA (Instrument Society of America)
- ISO (International Standards Organization)

The standardization of the FFBus has been promoted by the IEC TC 65 and ISA S50 Committees. To help expedite the development of the FFBus, the Interoperable Systems Project (ISP) was organized by Fisher Control, Rosemount, Siemens, and Yokogawa in August 1992. Later on, other companies organized committees to develop fieldbus. The process of FFBus standardization is depicted in Figure 2–7.

The open system interconnection (OSI) seven-layer reference model is the basis for most control system communication protocols. The ISA model is used as a framework to implement the network standards. The model's seven layers are physical, datalink, network, transport, session, presentation, and application. Figure 2–8 shows the OSI model. Some networks implement fewer levels, while the trend is to implement all the layers.

The present FFBus model has three parts (Figure 2–8):

1. Physical layer (OSI model layer 1)
2. Communication stack (OSI model layers 2 and 7)
3. User application (specified by FFBus)

The physical layer is defined by the IEC and ISA. The physical layer receives messages from the common stack and converts them into physical signals for transmission on the fieldbus medium and vice versa.

28 | INDUSTRIAL PROCESS CONTROL

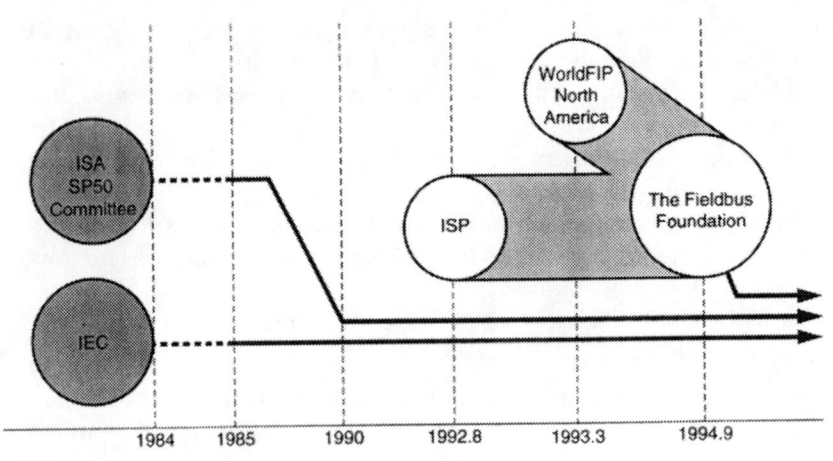

•1984
The standardization concept of digital communication protocol for field devices was proposed to IEC.

•1985
In IEC/TC65/SC65C, the new standardization work item was recognized and named Fieldbus.

•1990
The ISA SP50 Committee and IEC/TC65/SC65C/WG6 decided to collaborate on Fieldbus standardization.

•August, 1992
ISP was organized.

•March, 1993
WorldFIP was established.

•September, 1994
The ISP Association and WorldFIP North America were combined into The Fieldbus Foundation.
Since then, The Fieldbus Foundation has developed the internationally unified instrumentation specifications.
The Fieldbus standardization structure is configured by IEC, ISA, and The Fieldbus Foundation.

Figure 2-7 Process of foundation fieldbus standardization (courtesy of Yokogawa).

There are two FFBus specifications (Re IEC 1158-2 and ISA S50-02 standards): low speed (H1) and high speed (high-speed Ethernet [HSE]; originally called H2). The H1 FFBus has a transmission speed of 31.25 kbps, while the HSE's speed is 1 mbps or 2.5 mbps. The HSE FFBus can accommodate multi-component instruments and control units. Figure 2-9 shows H1 and HSE FFBus networks.

The H1 FFBus can use the wiring that was used by the 4–20 mA instruments. The H1 FFBus devices can be powered from the 4–20 mA cable. For the FFBus signaling, the transmitting device will deliver ±10 mA at 31.25 kbps into a 50 Ω resistor to create a 1.0-volt peak-to-peak, modulated on top of the DC supply voltage. The power supply can range from 9–32 VDC. For IS circuits, however, the power supply range depends on the employed IS barriers.

The H1 FFBus cable length depends on the number of devices and the wire size. The main cable run cannot exceed 1,900 m. The cable length is calcu-

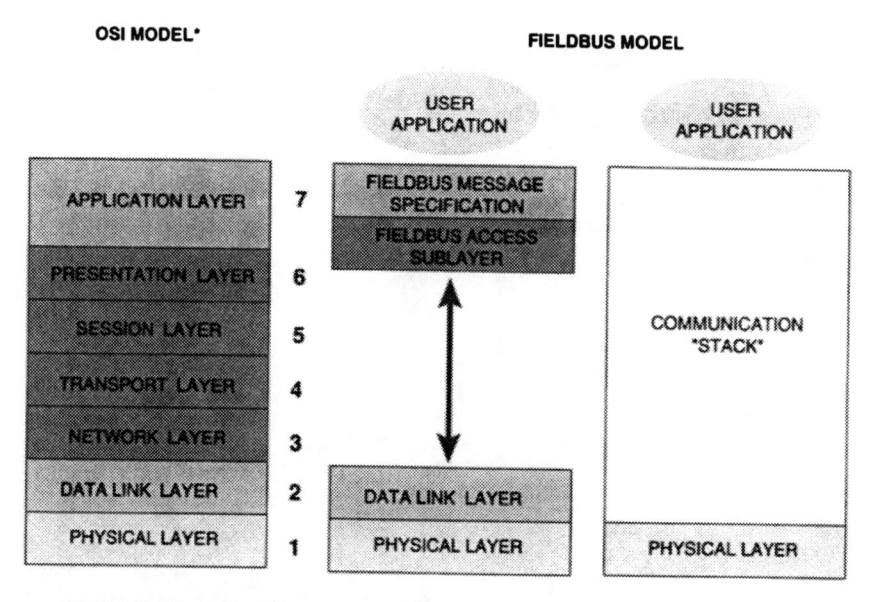

Figure 2–8 The OSI layered reference model.

lated by totaling the length of the trunk cable and all the spur lengths. The spur length depends on the number of instruments, although the maximum length is 120 m.

The configuration of control systems using FFBus is substantially different from those employing conventional analog instruments. The functional specifications for the SAS and field instruments should cover all the critical aspects of the FFBus, such as cable layouts, instrument layouts, system cabinets distribution, system interfacing, application engineering, documentation, testing, commissioning, and maintenance. Close cooperation between the contractor design team and the vendors is of paramount importance.

The FFBus instruments have extra parameters, which have to be accommodated. Parameters, such as process input filter time, compensation factor, ranges, device address, communication settings, and function block definitions, have to be established, programmed into the instruments, and properly recorded and documented. The SAS vendors offer special software/hardware packages to manage the parameters.

In the design of the SAS configuration, careful planning for the FFBus interfacing is necessary. This includes, for example, the number of buses, the number and type of branches, the quantity of devices per branch and per bus, the type of cables, and allocation of buses to control units. It is also critical to allow for future expansion.

The cable length and type for the buses and branches should be planned carefully. The following FB-1 cable type and the number of instruments per

30 | INDUSTRIAL PROCESS CONTROL

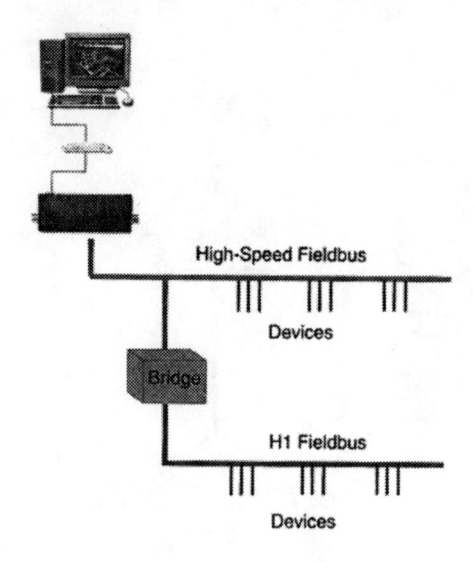

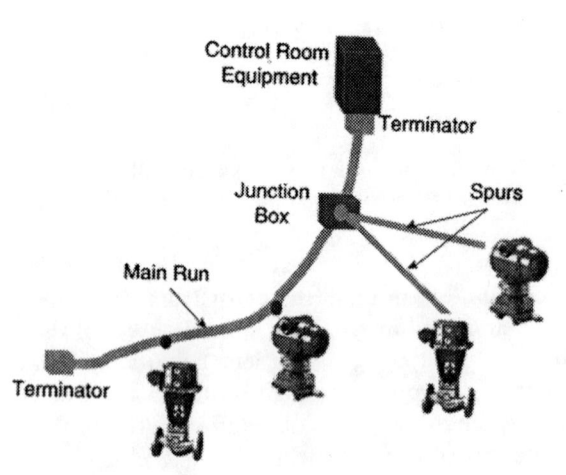

Figure 2-9 Foundation fieldbus low-speed and high-speed networks (courtesy of Fisher-Rosemount).

branch can be used as a guide (other factors, such as power supply and communication performance, may affect the number of devices):

- 1,900 m maximum for type A cable (twisted pair, individually shielded)
- 1,200 m maximum for type B cable (twisted pair, overall shielded)
- 200 m maximum for type C cable (not twisted pair, overall shielded)

For type A cables, the maximum length and the number of instruments per branch are:

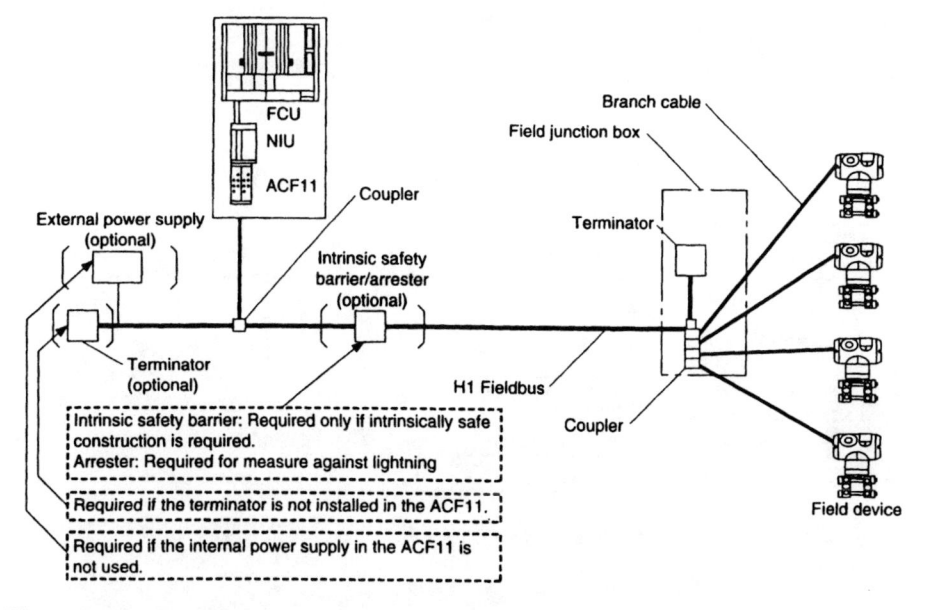

Figure 2-10 Foundation fieldbus star topology (courtesy of Yokogawa).

Maximum Length	Number of Instruments
120 m	1–12
90 m	13–18
30 m	19–24
0 m	25–32

During the design of control systems employing FFBus, the following subjects should be carefully planned:

- Cabling, terminations, junction boxes arrangement
- Transmission speed and sampling rates
- Address assignment to instruments
- Identifying data requirements for operation, maintenance, and configuration
- Spare requirements (for future expansion)

Two types of FFBus topology are available: star and bus types. Figures 2–10 and 2–11 show the star and bus topologies.

The type of topology chosen for a particular control system will obviously depend on instrument layouts and preferred roots for cables and cable trays. In both types of topologies, the limits on the cable lengths and the number of instruments per bus and the spare capacity for future expansion should be considered. FFBus cables, including intrinsic safety cables, should be routed separately from other cables.

The commissioning of control systems using FFBus needs different types of tools for testing, calibration, and fault analysis from those used for analog

32 | INDUSTRIAL PROCESS CONTROL

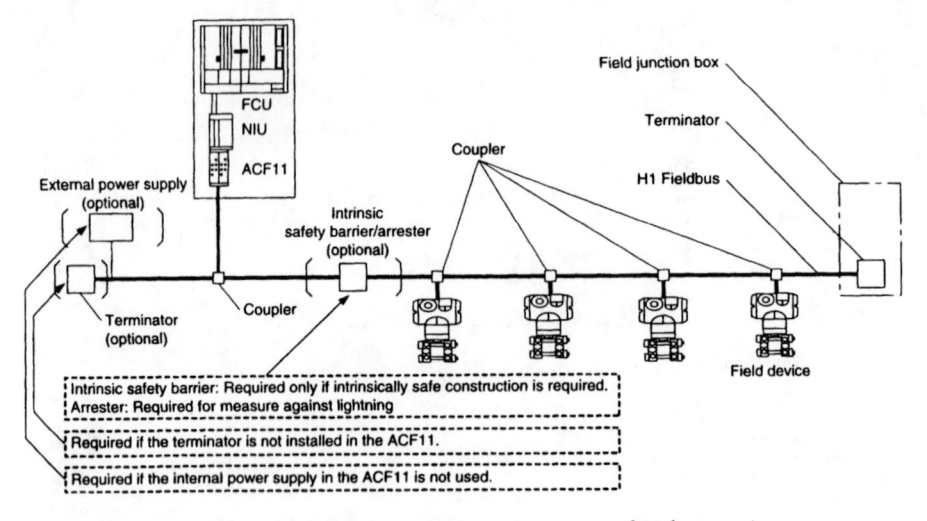

Figure 2-11 Foundation fieldbus bus topology (courtesy of Yokogawa).

instruments. Because the FFBus instruments are digital and yield much higher accuracy levels than analog instruments, specially designed calibration devices and signal analyzers are required. A bus analyzer is also needed to troubleshoot communication problems.

A critical aspect of control systems employing FFBus is communication. The communication system is far more complicated than the 4–20 mA analog system. Consequently, for the commissioning of control systems, control systems engineers are preferred to instrument engineers.

Another useful aspect of control systems employing FFBus is that the proportional integral derivative (PID) control functions can be implemented in the field instruments (transmitter or control valve). This approach has several advantages over the traditional control systems. The system will be more reliable, system response will be much faster, and the load on the DCS will be substantially reduced. This will allow the control unit to process advanced/multivariable control loops more efficiently. In traditional control systems, the DCS processing time on PID control loops and analog-to-digital conversion tasks is significant.

Savings in commissioning time and cost of control systems using FFBus versus traditional instruments will be substantial. The main factors contributing to this saving are as follows:

- Remote testing and calibration of field instruments (from CCR or LCR)
- Checking loop continuity without need for disconnecting any wiring
- Multiple instruments are connected to a single bus, which minimizes the quantity of wiring, cabling, terminations, and consequently reduces loop testing efforts

- High accuracy and stability of field instruments resulting from digitization, which reduces the possibility of deterioration in signal transmission and instrument accuracy
- No need to specify the instrument range or span for the FFBus instruments because they digitally measure the process parameters

Selection of the type of cable for FFBus is critical. Because several instruments are connected to a single bus and the signal strength may also be low, adherence to the individual instrument's wiring and termination requirements is critical.

In order to establish the acceptable bus length for a particular FFBus configuration, sevaral parameters must be considered. The following example will help with the calculation of the bus length.

Instrument/Cable Parameters
N = The number of field instruments = 5
V_{max} = 14 VDC
I_{max} = 250 mA
I_{inst} = 12 mA
V_{inst} = 9 VDC
R_{cable} = 24.6 Ω/km
Cable type = A (IEC 1158, ISA SP 50)

Calculations
Maximum allowed voltage loss = V_{max} − V_{inst} = 14 − 9 = 5 VDC
I_{total} = N × I_{inst} = 5 × 12 mA = 60 mA
Maximum allowed cable resistance = V_{max} ÷ I_{total} = 5 ÷ 0.06 = 83 Ω
Maximum allowed cable length = 83 ÷ 49.2 = 1.7 km

The cable length may also be restricted by maximum allowed capacitance. Similar calculations should be carried out for capacitance.

2.6 MULTIPHASE FLOW METERING

Multiphase flow metering (MPFM) is a young technology that has high potential, especially in the oil and gas industry. I specified the first true MPFM for offshore applications in 1993. Three MPFM systems were procured and installed for three satellite platforms. These MPFM units were used instead of test separators. As the MPFM vendors gain experience and improve the accuracy of the units, wide applications such as subsea flow metering, allocation metering, and perhaps fiscal metering will follow.

At present, most MPFM systems vendors are Norwegian. The Norwegian oil companies support the vendors by providing finance and allowing testing on offshore platforms. The results of online tests have been encouraging, and the installation of three Framo MPFM systems we procured in 1994 will encourage other oil companies to use them instead of test separators.

In a seminar in 1993,[6] the MPFM vendors indicated that MPFM systems suitable for subsea applications would be available by 1995, which seemed optimistic. By early 1998, many offshore oil/gas production platforms were provided with MPFMs (topside and subsea) instead of test separators. They include both topside and subsea applications. I have no doubt that MPFM will be adopted as the industry standard for oil/gas well testing (i.e., replacing test separators). It is also suitable for marginal fields, where the cost of traditional metering facilities may prove prohibitive.

Most of the MPFM systems I studied employ gamma-ray attenuation measurement and statistical methods (cross-correlation) to calculate phase fractions. A DP meter (e.g., venturi) and a homogenizer are also employed to measure total flow accurately. The future MPFM systems may use neural networks and pattern recognition to improve the accuracy of measurement.

Figure 2–12 shows the main components of the MPFM systems used in offshore projects (Framo, Bergen, Norway). The Framo Meter uses a mixer to yield a homogenous mixture, which is essential for accurate flow measurement.

2.7 SUBSEA CONTROL AND INSTRUMENTATION SYSTEMS

Subsea C&I systems can be divided into three main subsystems:

1. Subsea field instruments and control pods
2. Umbilical, carrying process variables, control signals, electric power, hydraulic power, and methanol
3. Topsides control units (master control station, electrical power unit, and hydraulic power unit)

Figure 2–13 depicts the main elements of a subsea C&I system. Because of the special subsea C&I requirements (environmental, signaling, reliability), design and engineering (conceptual, detail, commissioning) of subsea systems are carried out by subsea systems engineers, rather than by traditional instrument engineers. The following three sections describe the subsea C&I systems in detail to provide an overview.

2.7.1 Subsea Field Control and Instrumentation

Because of subsea environmental and application requirements, subsea C&I equipment is designed for the following special features:

- To withstand an underwater environment
- To require minimum maintenance and service
- To use minimum power
- To provide high reliability

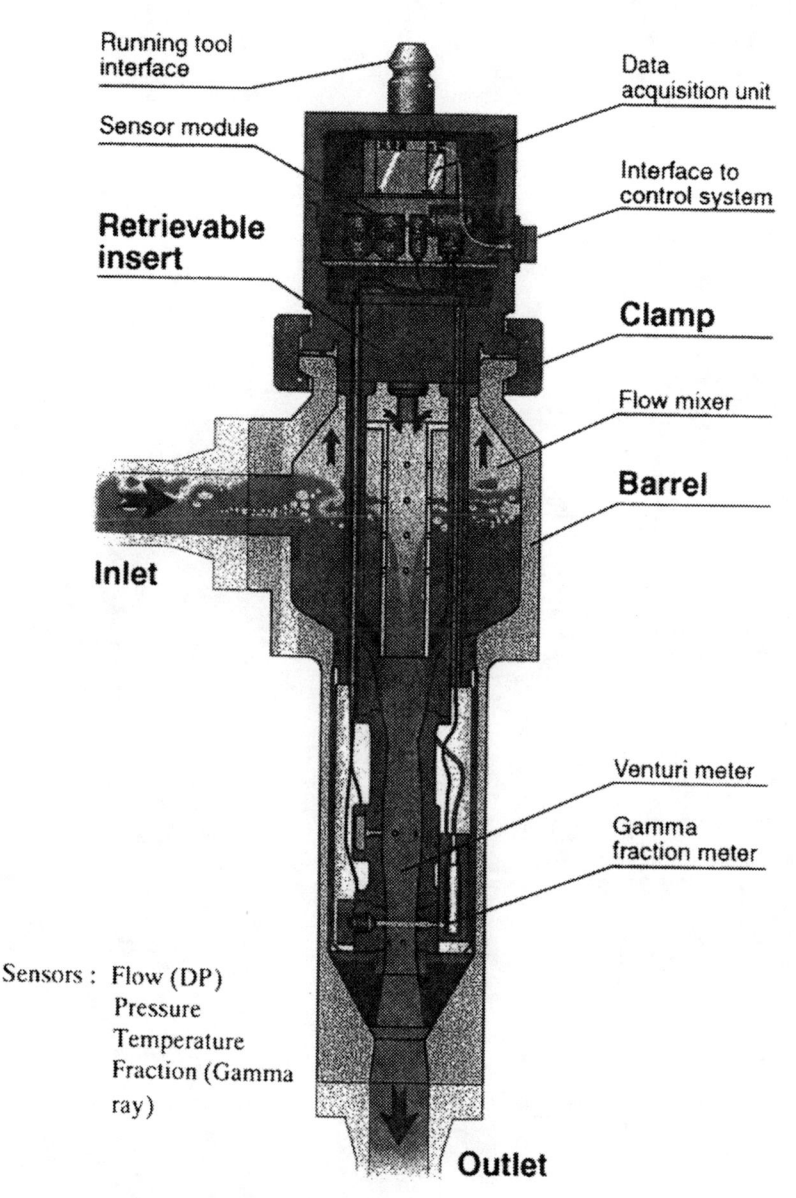

Figure 2–12   Framo multiphase flow metering system (courtesy of Framo).

In order to meet these requirements, a subsea field transmitter could cost 10 times its topsides counterpart. In addition, the subsea C&I systems are designed with a dual or triplicated configuration. The system should be failsafe and require minimum operator interaction for startup, shutdown, and normal operation.

The design and detail engineering of subsea C&I systems are carried out either by specialized engineering companies or by the subsea departments of large

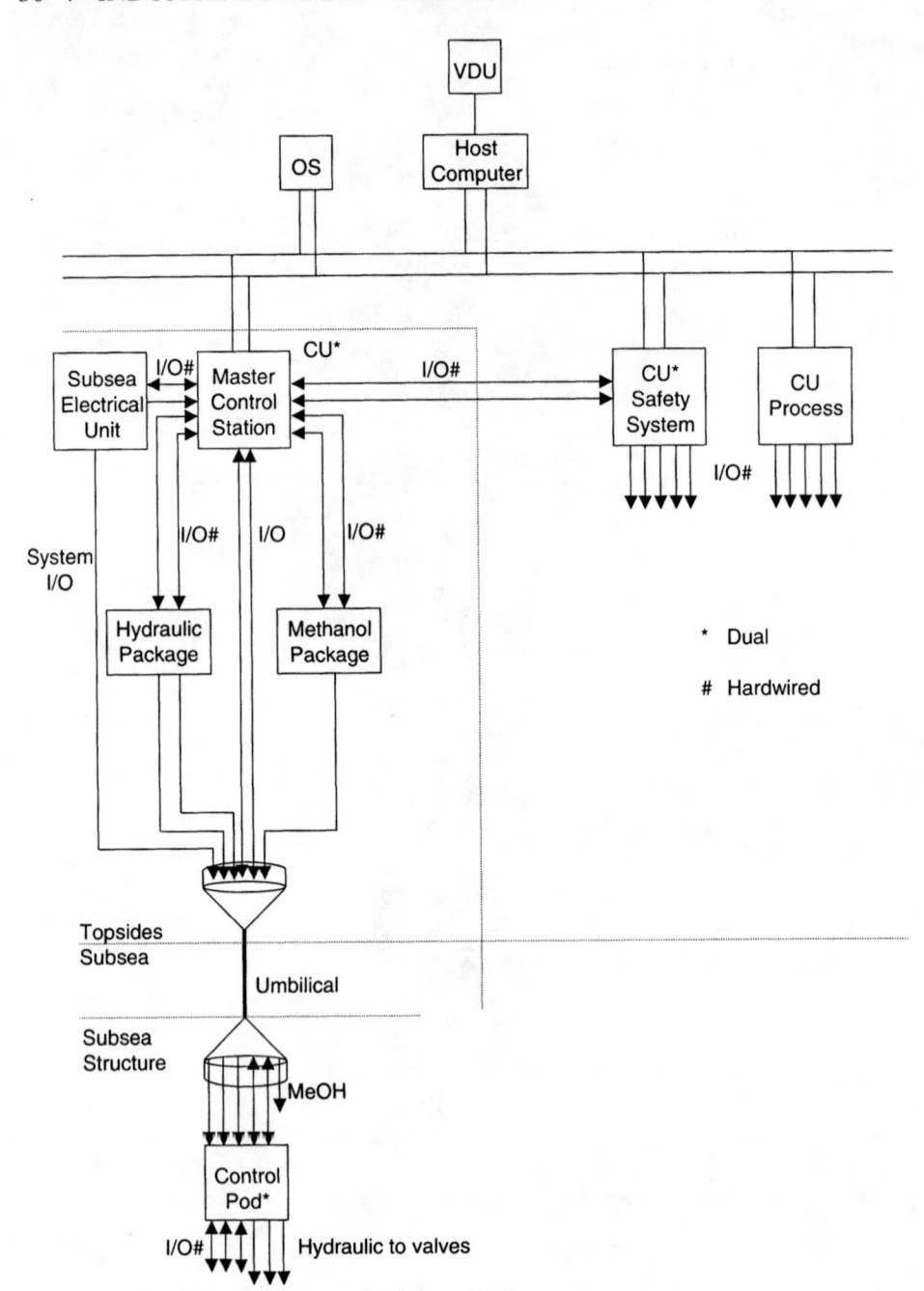

Figure 2–13 Subsea control and instrumentation system.

contracting/engineering firms. I prefer the latter arrangement because large companies can engineer/design the topsides and provide a smooth interface between subsea and topsides disciplines. This approach will not only benefit the logistics but will also improve management.

The subsea control units (electronics) are housed in control pods. Control pods interface with local subsea instruments via electrical and hydraulic media and topsides with the master control station (MCS) via umbilical (serial links). Power to control pods is provided via topsides hydraulic power units (HPUs) and electrical power units (EPUs). All power supplies (hydraulic and electrical) and transmission lines through umbilical are dual.

2.7.2 Umbilical

Current subsea systems can be provided with very long umbilicals, 50 km or more. This phenomenon, together with the possibility of using subsea MPFMs, is a decisive factor in the development of marginal offshore oil and gas fields. In a conference,[6] a speaker stated that the present uneconomical North Sea marginal fields capacity is as big as the currently producing fields.

Umbilical enters a distribution box on the platform, where the hydraulic pipes, methanol lines, electric, and process control signal cables will be branched to the methanol injection package, HPU, EPU, and MCS.

2.7.3 Master Control Station

The MCS is a dual microprocessor-based control unit that, in addition to controlling the subsea facilities, provides the following functions:

- Control of methanol system
- Electric power for subsea equipment
- Interface to SAS

A limited number of vendors offer MCS systems because of the specialized software (application) and hardware (interface to control pods) requirements. Normally, subsea system vendors procure hardware and software from a third party and then develop the application software (database) and various interfaces as required.

The normal operation and monitoring of the subsea process and the related facilities is via the platform PCS. The PCS operator station, located in the central control room (CCR), will be configured to provide all the necessary graphics, mimics, reports, and alarm displays. If the MCS belongs to the same family as the PCS hardware, the MCS–PCS interface will be via the DCS LAN; however, if the PCS and MCS hardware are different, gateways will be required for the two systems to interface. Of the many control systems I have evaluated, only a few offer a standard control unit, which can also be used for the subsea MCS. Examples of these are the Siemens SIMATIC PLC and ABB Advant Control Unit.

Where the MCS and PCS use the same hardware, interfacing the two systems is easy. In such systems, the interfacing problem is basically reduced to the development of a subsea process operation database in the PCS operator station. For systems with different hardware, however, all of the problems associated with serial links have to be dealt with. The ideal case would be where the platform PCS vendor is employed as the main control system vendor, and they employ a subcontractor to be responsible for the subsea control system.

2.8 VESSEL CONTROL SYSTEMS

2.8.1 General

In order to reduce the capital costs for the deepwater field oil/gas production projects, floating production, storage, and offloading (FPSO) vessels are used instead of fixed platforms. Such FPSOs require advanced control and monitoring systems for the vessel management, in addition to the normal production SAS. FPSOs may be subjected to violent natural forces, created by wind, waves, and currents. They require proper control to counteract these forces and keep the vessel position and heading within acceptable limits. Figure 2–14 shows the natural forces that act upon an FPSO and the resulting motions.

The vessel control systems (VCS) normally include the following subsystems:

- Dynamic positioning system
- Ballast control system
- Environmental, meteorological, and platform monitoring system

2.8.2 Dynamic Positioning

The dynamic positioning system (DPS) counteracts the forces (i.e., wind, waves, and currents) that naturally act upon the vessel and maintains the vessel within the desired position and heading. The DPS employs a mathematical model

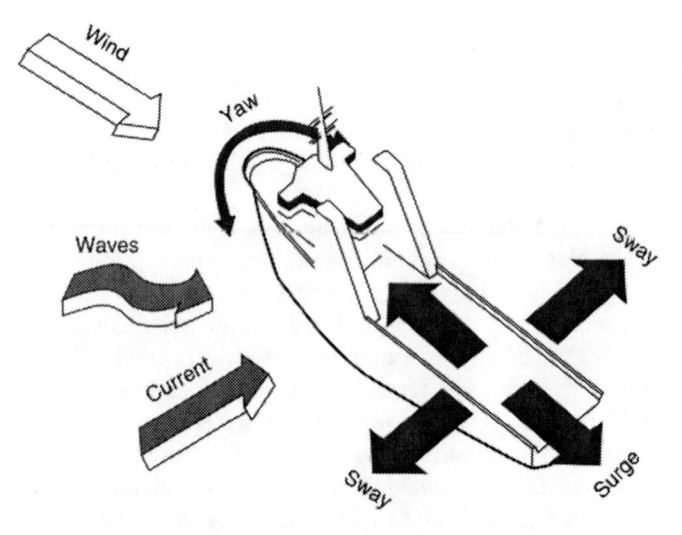

Figure 2–14 Forces acting on an FPSO and the resulting motions.

to estimate the sea currents and waves and the forces they impose on the vessel. The model is a hydrodynamic description of the vessel that calculates the vessel's reaction to the forces acting upon it.

Figure 2–15 shows the mathematical model used by the Simrad DP system.[7,8] The main outputs from the model are estimates of the vessel's position, heading, and speed in the three degrees of freedom: surge, sway, and yaw.

The DP controller produces an optimum output, which indicates the force required to apply to the vessel's thrusters/propellers in order to achieve the desired position/heading. The Simrad DP model minimizes fuel consumption and wear and tear on the vessel thrusters/propulsion system. The controller also allows manual control of the thrusters and propellers.

2.8.3 Ballast Control

The ballast control system (BCS) provides control and monitoring functions for the vessel ballast, bunker, bilge fluids, and various products/cargo and load/stability. The system controls and monitors routing and heeling by using relevant sensors (i.e., level, temperature, pressure) and start/stop of pumps and open/close of valves.

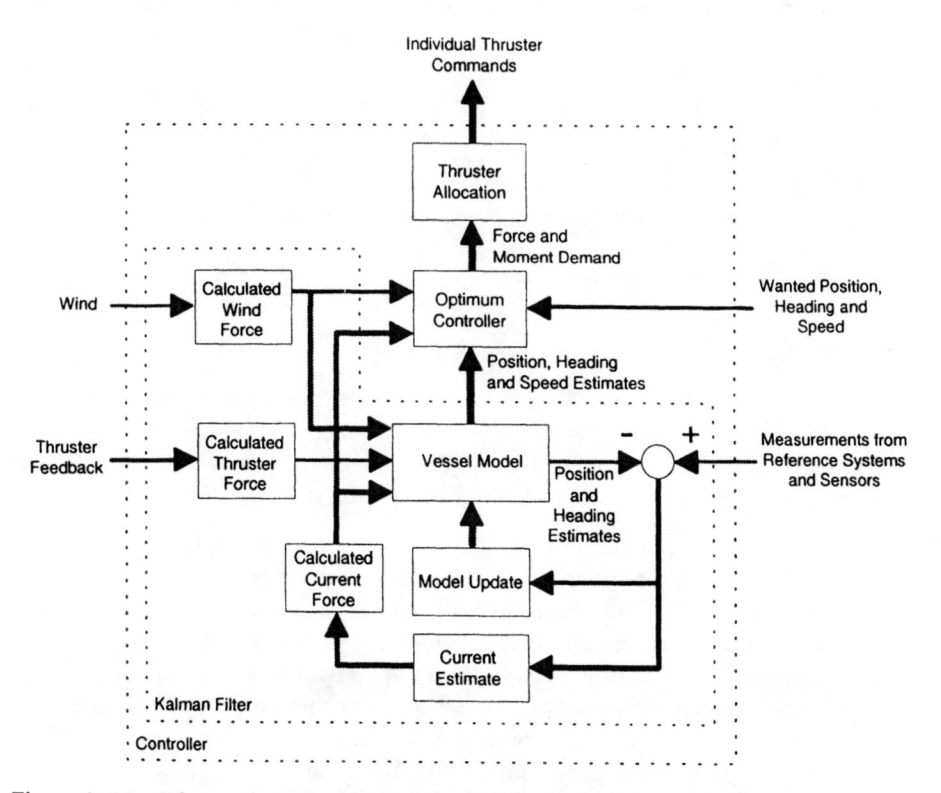

Figure 2–15 The mathematical model used by DPS (courtesy of Simrad).

2.8.4 *Environmental, Meteorological, and Platform Monitoring*

Environmental, meteorological, and platform (EMP) monitoring systems provide data acquisition and monitoring (displays, alarms, and reports) for the following parameters:

- Wind speed and direction
- Sea current speed and direction
- Wave height and period
- Water depth
- Sea temperature
- Air pressure, temperature, and humidity
- Platform roll, pitch, heave, surge, sway, yaw
- Platform line tension and length
- Strain gauges

2.8.5 *Operator Interface*

In integrated SAS, the VCS operator station uses the same hardware and software (operating system) as the process control operator stations; however, special functions for vessel control and operation require extra facilities, such as three-axis joystick control, trackerball, dedicated keys and lamps, heading control, and a simulator.

The operator simulator is a valuable tool that will assist the operator in the following tasks:

- Training
- Predicting the effect of changes in weather, waves, line breaks, and so on
- Planning new strategies

2.9 CONDITION MONITORING

2.9.1 *General*

Condition monitoring of process equipment is not a new subject, but two main factors are the driving force behind its ever-increasing use:

1. Powerful computers and the use of intelligent systems (neural networks) and mathematical models (Markov model, reliability–availability analysis) allow online monitoring of some useful parameters, such as vibration, movement (misalignment), and equipment wear (lubricating oil contamination). Some vibration signals and pattern recognition systems require very fast sampling (several scans per millisecond). Such measurements have just recently become possible. They require specialized computers, which are presently outside the capability of process control systems.

2. The costs of process equipment maintenance and losses caused by unscheduled shutdowns are substantial. The cost of annual maintenance in the UK process industry is well over £20 billion. Reductions in such costs/losses are the prime objective of all plant managers.

2.9.2 Maintenance Strategies

Several maintenance strategies are available to plant managers, namely:

- Corrective maintenance
- Preventive maintenance
- Predictive maintenance
- Detective maintenance

Corrective maintenance is the simplest method by which repair is carried out after a failure has occurred. Obviously, if the failed item of equipment is a critical part of the plant, or if it requires substantial repair efforts, corrective maintenance will not be a suitable choice.

Preventive maintenance requires regular maintenance in order to keep troublesome failure modes at bay. In a preventive maintenance strategy, we assume that regular maintenance will prevent major equipment failures and plant shutdowns.

In *predictive maintenance*, the system monitors several relevant equipment condition parameters, such as vibration, lubricating oil, and temperature, in order to infer the status of the plant facilities. Condition monitoring of rotary equipment, such as pumps, compressors, generators, mixers, and paper machines, can reveal valuable information. The analysis of such data can predict the equipment failure well before the failure occurs. Condition monitoring techniques use the latest analytical methods, such as pattern recognition and neural networks, in order to assess the health of critical process machinery (e.g., generators, compressors, and large pumps). Such analysis will reveal the weak equipment components and prominent failures. Detecting weak components, replacing them, and preventing unscheduled shutdowns can save substantial costs and enhance the useful life of critical equipment.

Detective maintenance applies to devices that only need to work when there is a demand on them (e.g., ESD valves). Shutdown valves may not need to operate for months, or even years. They require periodic checks to ensure that they will operate satisfactorily when there is a demand on them.

2.10 RELIABILITY AND AVAILABILITY ANALYSIS

2.10.1 Introduction

There are numerous articles and reports on the subject of reliability and availability, however, most of them cover only a small aspect of the subject. This section provides readers with some useful and relevant information based on the latest developments in reliability analysis, which will help with the design of C&I systems. The subject is introduced in a simple, straightforward manner by using examples from C&I systems. If readers are interested in more mathematical analysis, reference should be made to works by Churchley, Goble, and Hellyer.[9-11]

Readers are encouraged to apply the suggested analysis to the C&I systems during the design of such systems. In the specification of requirements of a large integrated control and safety system, I included a section highlighting reliability requirements. The vendor submitted the reliability analysis document for my review and approval. The document had no resemblance to my specified requirements. I rejected the document and insisted that the vendor should submit the document according to the specified standards (various SINTEF standards).

The vendor reliability engineer made various excuses (e.g., other clients are satisfied with the offered document, they are not familiar with SINTEF standards, some information cannot be released, some information is not available, etc.) in order to attempt to change my mind. I requested a meeting with the engineer and explained my requirements in detail. When the engineer fully understood the requirements, he felt able to comply.

2.10.2 Analysis

To apply reliability analysis to the design and configuration of C&I systems, we do not need to delve into complicated mathematical methods. Indeed, only basic calculations and simple routines are required. Pictorial representation of the analysis will help with understanding and decision making regarding the selection and configuration of C&I system elements. By using this simple analysis, a highly reliable and fault-tolerant configuration can be designed, where mean time between failures (MTBF) of several decades or hundreds of years can be attained. Unfortunately, some C&I engineers still believe that the old-fashioned relay-type, solid-state systems or pneumatic instruments are more reliable than modern microprocessor-based systems.

There are several ways to improve the reliability of C&I systems, as explained as follows:

- Use of redundancy (1oo2, 2oo2, 2oo3) for systems
- Use of derating for components
- Use of self-checking (periodic testing)
- Inclusion of field instruments in the analysis
- Application of quality assurance/quality control (QA/QC) and thorough testing (design review, software review, module testing, integration testing, heat and soak testing)
- Use of full automation and fool-proofing techniques
- Use of preventive maintenance
- Well-designed unit and system configuration

Implementation of these principles will increase the cost of a C&I system substantially, depending on how far each method is applied. Figure 2-16 shows the optimum reliability-cost analysis for a typical C&I system.

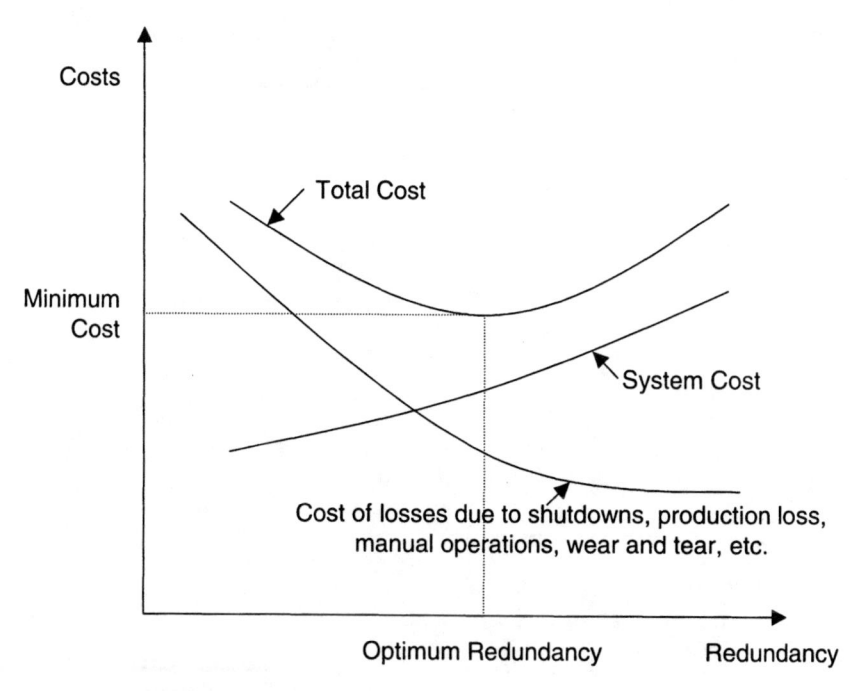

Figure 2-16 Reliability–cost analysis.

The use of redundancy (1oo2, 2oo2, 2oo3) for safety systems (ESD, F&G) is well understood, and all systems usually incorporate it into their design. The 1oo2 redundancy is suitable for systems in which safety is more critical than loss of production, whereas 2oo2 is the reverse. A compromise is 2oo3, where both safety and production are improved. Tables 2-1 and 2-2 show the probability of failure values for a single, 1oo2, and 2oo3 configuration for a CU and process transmitter, respectively. MTBF is assumed to be 5 years for the CU and 10 years for the transmitter. The following equations are used:

$P(1oo1) = 1 - \exp(-t/MTBF)$

$P(1oo2) = (1 - \exp(-t/MTBF))^2$

$P(2oo3) = 1 - 3\exp-(-2t/MTBF) + 2\exp(-3t/MTBF)$

It is obvious from these reliability figures that as the system age increases, the 2oo3 system becomes less reliable (than 1oo1). Readers should not be alarmed because these reliability calculations are purely mathematical and do not apply to real-life 2oo3 redundant systems. These calculations assume that the suggestions to improve system reliability as indicated earlier in this section are not incorporated. By using the aforementioned principles, we can increase the reliability of a 2oo3 system significantly (by a factor of 5 to 10).

Table 2-1 Probability of Failure Values for a Control Unit

Years	P (1oo1)	P (1oo2)	P (2oo3)
0.5	0.095	0.009	0.025
1	0.181	0.033	0.087
1.5	0.259	0.067	0.167
2	0.330	0.109	0.254
2.5	0.393	0.155	0.343
3	0.451	0.204	0.427
4	0.551	0.303	0.576
5	0.632	0.400	0.694
6	0.699	0.488	0.782
7	0.753	0.568	0.848
8	0.798	0.637	0.894
9	0.835	0.697	0.927
10	0.865	0.748	0.950

Table 2-2 Probability of Failure Values for a Transmitter

Years	P (1oo1)	P (1oo2)	P (2oo3)
0.5	0.049	0.002	0.007
1	0.095	0.009	0.025
1.5	0.139	0.019	0.053
2	0.181	0.033	0.087
2.5	0.221	0.049	0.125
3	0.259	0.067	0.167
4	0.330	0.109	0.254
5	0.393	0.155	0.343
6	0.451	0.204	0.427
7	0.503	0.253	0.505
8	0.551	0.303	0.576
9	0.593	0.352	0.639
10	0.632	0.400	0.694

Readers may ask why I have shown these tables if they are not relevant to real-life systems. There are two reasons for this: (1) these figures are quoted by those vendors who do not offer a 2oo3 system; and (2) the values apply if the methods to improve reliability are not implemented (i.e., redundant systems that are not specifically designed for safety systems).

The 2oo3 system's reliability can be improved substantially if common causes of failures (i.e., those sources that cause the failure of more than one component) can be reduced significantly. The following sources are the main causes of common mode failures.[10,12]

Source of Failure	Percentage of Failure
Latent design errors	70–90%
Operational errors	50–75%
Maintenance errors	50–75%
Abnormal environmental stress	10–30%
Wearout	1–5%

To reduce the design errors, we have to implement quality assurance/quality control (QA/QC) and thorough testing, and a well-designed system configuration, which were described earlier. The more robust the system configuration and the more efficient the QA/QC and system testing, the lower the chance of design errors yielding common failures will be. To minimize operational and maintenance errors, full automation, fool-proofing techniques, and preventive/planned maintenance should be employed. Derating and a well-designed system (units, configuration) will substantially reduce the effect of environmental stresses. Of course, a well-designed environment (control rooms) is always necessary.

I recommend the following strategy to combat wearout causes. The vendors of control and safety systems continually upgrade their system's components (e.g., I/O cards, CPU, RAM). There are two reasons for this upgrade: (1) to enhance the system's capability, and (2) to improve the design and eliminate problems identified as a result of feedback from customers. Vendors normally offer incentives to their customers for component upgrade. End-users should accept this offer and replace their installed components. This strategy will benefit both vendors and customers as follows:

- Reduction in wearout effects
- Improved system performance
- Lower spare parts burden (for vendor)

It is important to include field instruments (transmitters, valves) in the reliability analysis because it is common knowledge that between 60 to 80 percent of shutdowns are caused by the failure of these items. There is no point in using a highly reliable ESD system if some single process transmitters can cause a major hazard or a total plant shutdown. Although HAZOP studies help reduce the effect of such shortcomings, very few control/safety systems reliability analyses include field instruments. Figure 2–17 shows the system block diagram for reliability analysis.

If any of the system components shown in Figure 2–17 fails, the system will fail, which will cause a shutdown. Some components (e.g., power supply, RAM,

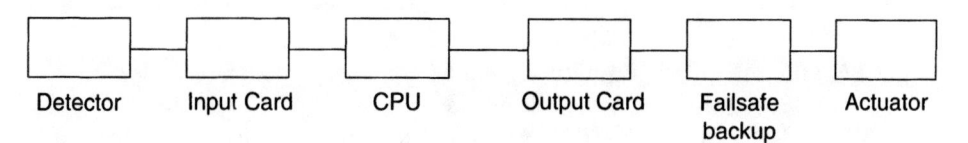

Detector — Input Card — CPU — Output Card — Failsafe backup — Actuator

Figure 2–17 Reliability analysis system block diagram.

LAN cards) are not shown. These should be included in the analysis. If some elements are redundant (1oo2, 2oo2, 2oo3), these should be shown as parallel items. The calculation methods and the MTBF for some components, which are not readily available, should be made according to SINTEF standards (see Section 2.10.3).

One main advantage of 2oo3 systems (versus 1oo1, 1oo2, and 2oo2), which most engineers tend to ignore, are their superior availability. Availability depends on MTBF and mean time to repair (MTTR).

Availability = MTBF ÷ (MTBF + MTTR)

The formal definition of availability does not reveal the superiority of 2oo3 redundancy. In a 1oo1 or 1oo2 configuration, a single failure causes a shutdown and requires immediate maintenance attendance. In a well-designed 2oo3 safety system, however, a single failure can be tolerated for several days to several months, depending on the system's robustness. This is especially useful in cases where a shutdown can cause total production loss or where remote unstaffed plants are involved. For example, an offshore installation may include several unstaffed satellite platforms and/or some subsea templates. A shutdown of subsea wells is very costly, and servicing satellite platforms is also expensive and may not be possible at short notice. If a satellite platform is shut down at night or during stormy conditions, long production losses may be inevitable.

2.10.3 Standards

All control and safety system specifications include various standards and codes of practice to cover the reliability requirements. In most cases, however, I believe that both vendors and system engineers do not pay adequate attention to the stated requirements. It is imperative that all of the stated standards are clearly adhered to. The reliability analysis documents produced by vendors will be useless unless they clearly indicate the weak points of the system and suggest how these can be eliminated. The QA/QC aspects of the system hardware and software should be thoroughly reviewed, and it would be beneficial if an independent company audits the vendor. The following is a list of relevant standards that I find adequate for C&I systems:

BS 5750	Quality systems
BS 5887	Code of practice for the testing of computer-based systems
IEC 409	Guides for inclusion of reliability clauses into specification for components (or parts) for electronic equipment
IEC 605	Equipment reliability testing
ISO 9000	Parts 1, 2, and 3
SINTEF SFT75 F89025	A88011 Reliability of safety shutdown systems Reliability data for computer-based process safety systems

F89023	Reliability prediction handbook; computer-based process safety systems
F88035	Reliability and availability of computer-based process safety systems
F90002	Guidelines for specification and design of process safety systems

2.11 SAFETY INTEGRITY LEVEL AND IEC 61508

2.11.1 Introduction

This section introduces the principles of the IEC 61508 standard in a brief, understandable discussion. The purpose of installing a safety system in a plant is to reduce the risks posed by the process equipment and its control system. Although it is preferable to provide an independent safety system, a combined control and safety system is acceptable. In the case of small plants (e.g., an offshore satellite platform), a combined SAS is more appropriate. Combining the safety and control systems not only reduces the cost and space but also significantly improves the reliability of the control system.

The IEC 61508 suggests guidance and recommendations on good practice for the design and implementation of safety systems, in order to reduce risks posed by process and control equipment. Recommendations are broad and cover planning, management, documentation, assessment/analysis, hardware, software, and all safety-related activities. The whole life cycle of the system—from conceptual design to specifying the system requirements, hazard and risk analysis, testing, validation, installation, commissioning, operation, maintenance, repair, and disposal—is covered by the standard.

The IEC 61508 is in seven parts:

Part 1: General requirements (management, documentation, technical requirements)
Part 2: Requirements for electrical/electronic/programmable electronic safety-related systems
Part 3: Requirements for electrical/electronic/programmable electronic hardware/software
Part 4: Definitions, abbreviations
Part 5: Examples of methods for determining safety integrity levels
Part 6: Guidelines on the application of Parts 2 and 3
Part 7: Overview of techniques and measures and references to further information

2.11.2 Risk Analysis

A fundamental activity in the design of a safety system is the analysis of the risks imposed by the equipment under control (EUC) and its control system. This will include hazard identification, hazard analysis, and risk assessment. This activity requires the assistance of a team, whose members have complementary experience in the process. A hazard and operability (HAZOP) study will be effective. There are two aspects of risk, namely (1) the probability of it happening and (2) the potential consequence if it does.

In many cases of system failures, the probability can be calculated by using numerical values based on past experience and estimated figures from reliable sources. If the numerical values are not available or not reliable enough, however, the hazards must be analyzed qualitatively.

The likelihood of failure occurrence and the consequences are ranked as follows:

Categories of Likelihood	Definition	Failures per Year
Frequent	Many times in system life	$>10^{-3}$
Probable	Several times in system life	10^{-4} to 10^{-3}
Occasional	One time in system life	10^{-5} to 10^{-4}
Remote	Unlikely in system life	10^{-6} to 10^{-5}
Impossible	Very unlikely in system life	10^{-7} to 10^{-6}
Incredible	Cannot believe it could happen	$<10^{-7}$

Categories of Consequences	Definition
Catastrophic	Multiple loss of life
Critical	Loss of a single life
Marginal	Major injuries to one or more persons
Negligible	Minor injuries at worst

Risk levels and classifications are shown in the following table:

Likelihood	Catastrophic	Critical	Marginal	Negligible
Frequent	I	I	I	II
Probable	I	I	II	III
Occasional	I	II	III	III
Remote	II	III	III	IV
Impossible	III	III	IV	IV
Incredible	IV	IV	IV	IV

Interpretation of risk classes is shown in the following table:

Risk Class	Interpretation
I	Intolerable Risk
II	Undesirable Risk
III	Tolerable Risk
IV	Negligible Risk

It is obvious that the highest risk is the cell representing frequent–catastrophic and the lowest is the cell represented by incredible–negligible. The ALARP (as low as reasonably practicable) principle can be used to define *tolerability*. Tolerability is industry dependent. For example, a tolerable risk in a pulp and paper mill may be undesirable in an oil production plant or intolerable in an atomic power plant.

A Class II risk may be tolerable if risk reduction is not practicable or the cost of risk reduction is grossly disproportionate to the gained improvement. A Class III risk may be tolerable if the cost of risk reduction would exceed the gained improvement. A Class IV risk is acceptable, although it may need monitoring.

2.11.3 Safety Requirements

The risk levels and classification tables highlight how risks may be reduced. For instance, a Class I risk in the cell probable–critical may be reduced to Class II by taking the following actions to reduce its likelihood to occasional, or to reduce its consequence to marginal.

Safety requirements statements can be made based on which risks should be reduced and how they could be reduced. Clearly it is preferable to (1) eliminate the risk (or reduce its likelihood as far as is practicable), (2) mitigate its consequences, and (3) install emergency systems. Each safety requirement must have two elements: a safety function and an associated safety integrity level. Safety functions, which will reduce risks, must be allocated to safety-related systems.

The safety-related systems—process shutdown system (PSD), emergency shutdown system (ESD), blowdown system (BD), fire and gas system (F&G), and high-integrity process protection system (HIPPS)—are specified, designed, and installed to help reduce risks. The detailed requirements of these systems are covered by IEC 61508 standard. Other means of risk reduction (such as using relief valves, dual redundant PCS, highly qualified and experienced operators and maintenance crew) should also be considered to reduce risk. Control systems engineers, working in specific industries, should prepare brief and concise specifications indicating the safety system requirements.

2.11.4 Safety Integrity Levels

In Part 4 of the IEC 61508 standard, *safety integrity* is defined as "the likelihood of a safety-related system satisfactorily performing the safety functions under all the stated conditions, within a period of time," and a *safety integrity level* is defined as "a discrete level (one of four) for specifying the safety integrity requirements of safety functions."

Safety integrity levels (SILs) are derived from the assessment of risks, although they are not a measure of risk. They are a measure of the required reliability of a system or function. They relate to the target reliability of dangerous failures of the system in question. In general, the greater the required risk reduction, the more reliable the safety-related system, which means the higher the SIL level.

The following table shows the SIL levels and the correspondent probability of failure to perform safety functions on demand and for continuous/high-demand cases.

Safety integrity level	Probability of failure on demand (low demand)	Frequency of dangerous failures per hour (continuous/high demand)
1	$\geq 10^{-2}$ to 10^{-1}	$\geq 10^{-6}$ to 10^{-5}
2	$\geq 10^{-3}$ to 10^{-2}	$\geq 10^{-7}$ to 10^{-6}
3	$\geq 10^{-4}$ to 10^{-3}	$\geq 10^{-8}$ to 10^{-7}
4	$\geq 10^{-5}$ to 10^{-4}	$\geq 10^{-9}$ to 10^{-8}

The relation between SIL levels and DIN/TUV (a certifying authority) levels are as follows:

- SIL 0 = TUV AK 1
- SIL 1 = TUV AK 2/3
- SIL 2 = TUV AK 4
- SIL 3 = TUV AK 5/6
- SIL 4 = TUV AK 7/8

2.11.5 A Practical Approach

The IEC 61508 is not easy to comprehend or specific in giving instructions, nor does it indicate the required measures for assessing rules to apply to the design and configuration of safety systems. It gives general guidance on the development and management of safety-related systems throughout their life, from conception to decommissioning.

In my view, more specific and easier to apply instructions are required for specifying, designing, developing, and configuring safety systems. In a large offshore oil and gas production platform or FPSO, the safety-related systems include PSD, ESD, BD, F&G, and HIPPS. The primary plant control and monitoring system is PCS; however, the PCS is normally used for monitoring and operation of the total plant, which includes the safety systems. Hence consideration should be given to the reliability of the PCS. If the operator interface, which is an integral part of the PCS, fails, then the plant will be shut down after a preset time, even though the safety systems may be healthy.

For large installations, PLCs are used for PSD, ESD, BD, F&G, and solid-state systems for HIPPS. Depending on the number of inputs/outputs, the following system configurations may be employed:

1. A combined PSD, BD, and ESD system
   A dedicated F&G system
   An independent HIPPS

2. A combined PSD and low-level BD/ESD system
   A dedicated F&G system
   A combined HIPPS and high-level BD/ESD system
3. A dedicated PSD system
   A dedicated BD/ESD system
   A dedicated F&G system
   An independent HIPPS
4. Several distributed and combined PSD, BD, ESD, and F&G systems
   An independent HIPPS
5. A combined control and safety system; this is only applicable for small plants (e.g., a satellite oil and gas production platform)
   An independent HIPPS

For a combined system, the total system must be designed to meet the highest required SIL. The immediate benefit of installing a combined system is that the systems with lower SILs are upgraded to the highest SIL. The primary objective of combining safety systems is cost reduction, although upgrading safety (control) systems with lower SILs should not be underestimated.

Following the design of process flow diagrams (PFDs) and P&IDs, the process equipment and the safety-related equipment (e.g., transmitters, switches, valves, ESD, F&G) will be carefully studied to allocate SILs. This activity requires a thorough analysis and assistance from experienced engineers in several disciplines (e.g., process, safety, operation, control systems, and some package vendors). During the study of safety-related systems' inputs/outputs for SIL allocation, more care should be taken for parameters with higher SILs.

To help with the system configuration during the early stages of a project (i.e., front-end engineering, conceptual design, early detail design), where adequate data for SIL analysis is not available, the following table may be used:

Safety System	SIL Level
HIPPS	4
ESD/BD (high-level)	3
ESD/BD (low-level)	2
F&G	2
PSD	1

It is also recommended to specify the PCS with dual-redundant control units. This will not only reduce the possibility of spurious shutdowns but will also place less demand on safety systems.

All SAS vendors offer safety systems certified to SIL 3. The safety control units use 1oo2D, 2oo3 (triple modular redundancy [TMR]), or 2oo4D (quadruple modular redundancy [QMR]) configurations. Where high availability is required (e.g., in remote satellite platforms), the TMR and QMR systems are superior to 1oo2D. In normal applications, where a system fault can be rectified readily (1–2 hours), a 1oo2D system may be preferred.

3

Systems Theory

3.1 INTRODUCTION

Before describing the systems theory, it will be useful to explain what we mean by a "system." A system is the collection of elements within a defined boundary, which are interconnected in a specific fashion, to perform a specific set of tasks. Figure 3-1 shows a typical simple system.

The "interconnections" of the elements of the system are as important as the "elements" themselves, as will be seen later on. A system can be as complicated as the body of a human being (by far the most complicated system) or as simple as a plug used to connect our TVs, radios, PCs, and the like to the electricity supply sockets mounted in our homes or office walls. Let us study the elements, their interconnections, and the interfaces of the simple plug.

Plug Elements	Body
	Three pins (live, neutral, earth)
	Fuse
	Fuse seat
Element Interconnections	Pins–Body
	Fuse–Pins (live and neutral)
	Fuse seat–Body
Plug Interfaces	Socket
	Electricity supply
	Appliance lead
	Environment (room, office, garden, warehouse)

If we think harder or consider more sophisticated plugs, we may add extra elements, interconnections, or interfaces to this list. As readers can appreciate, each element, interconnection, and interface is critical and must be designed correctly, otherwise the plug is useless. For example, if the fuse or the pins are a wrong size, the plug may not fit in the socket or may create hazards for the equipment or the user. It is obvious that a PCS, ESD, or F&G system may have thousands of elements, interconnections, and interfaces. Therefore, the control

53

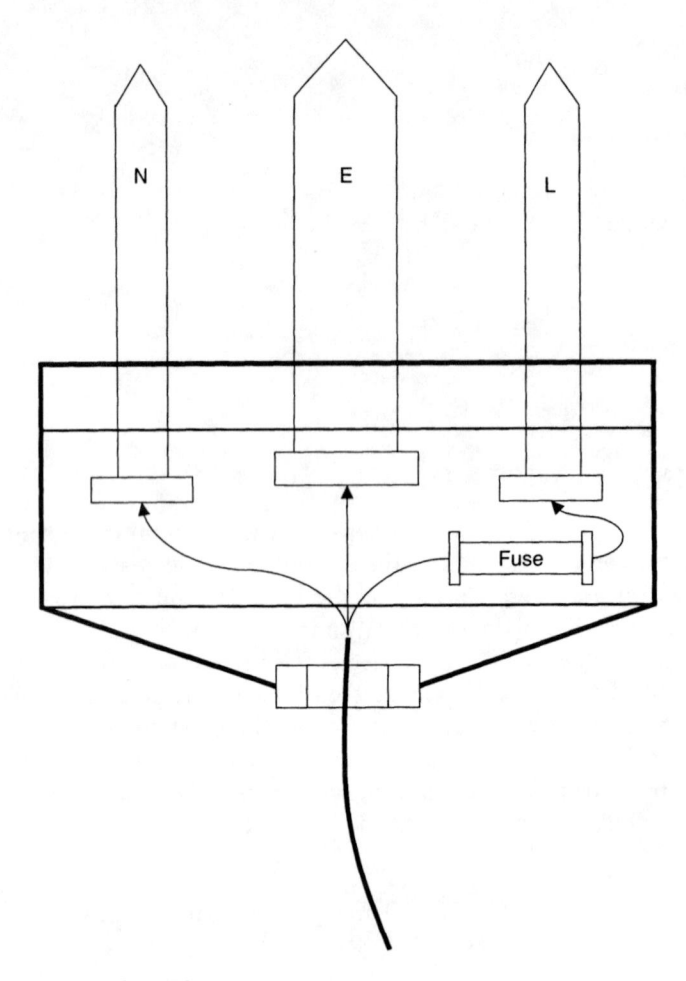

Figure 3–1 A typical simple system.

systems engineer and the vendor must ensure that all of the system elements, their interconnections, and system interfaces are designed properly.

From an engineering viewpoint, systems can be divided into two broad types: business systems and control systems. A system comprising a mainframe (or several minicomputers) and hundreds or thousands of terminals (spread over a large geographic area, e.g., a country) interlinked via a communication network (LAN, WAN, MAN), controlling and monitoring the operation of a large organization (e.g., a national or international bank) is a business system. A control system may consist of many control units, operator stations, a host computer, and field instruments (sensors, transmitters, valves) interconnected via a LAN, fieldbus, or the like.

Business systems are normally concerned with offline data, whereas control systems deal with real-time parameters. Fast scanning and processing (millisecond) of data are necessary in order to efficiently control and monitor the operation of process plants. Engineers who design business systems are called systems engineers, whereas those designing process control systems are control systems engineers. The real-time feature is what makes the process control systems complicated and requires the application of three extra theories in addition to the systems theory (i.e., control theory, sampling theory, and information theory).

Systems theory, in simple terminology, states that in designing a system, not only every element of the system has to be specified correctly, but all the interfaces between the elements of the system and between the system and the outside world have to be designed properly too. The elements of a system have a unique interrelationship between each other. The status, characteristics, and behavior of the elements of a control system at any moment in time depend on other elements of the system and the outside systems interfacing with the system.

Although this explanation seems simple and understandable to any engineer, I must stress that it is far from it. It is best to clarify the statement by an example. Learning systems theory is similar to learning to drive a car. To become a driver, one needs instructions and practice (i.e., experience) and to be fully conversant with the highway code (i.e., theory). Control systems engineering is a highly specialized job that requires both theoretical and practical knowledge. Learning the systems theory without having relevant experience is similar to learning to drive a car without any practice or driving instructions. Experience in systems engineering without understanding the systems theory is similar to driving a car without knowing the highway code.

The theoretical aspect of control systems engineering is the critical part, for two reasons: (1) control systems engineering is a new field, and few universities offer such a course; and (2) the subject is so broad that only the most experienced and knowledgeable engineers will appreciate what is involved. A typical control system consists of the following:

System Hardware	Electronics (control unit, operator station, etc.)
	Power supplies
	Printers
	Racks, cabinets, cables, etc.
System Software	Operating system
	Application software/firmware (control algorithms, etc.)
	Database (ladder logic, I/O conditioning, etc.)
	Utilities
Internal Interfaces	System LAN (highway)
	I/O bus
	Serial/Parallel links to peripherals

External Interfaces   Process interface (aliasing phenomenon)
Subsystems interface (gateways)
Foreign systems interface (serial links)
Operator interface (displays, alarms, etc)
Engineer interface (configuration, maintenance)
Management interface (reports, archiving)

It is almost impossible to find an engineer who is thoroughly experienced and familiar with all aspects of a control system as indicated in this list. Such an engineer should be conversant with the following fields:

- Control and Instrumentation Engineering
- Computer Engineering
- Electronics Engineering
- Electrical Engineering
- Process Engineering
- Communications Engineering

It is not normally possible for an engineer to learn all of these subjects in degree courses; however, it is practical to study three to four subjects during degree courses (i.e., BS and one or two MS courses) and learn the others during a working life.

3.2 OTHER THEORIES

3.2.1 Sampling Theory

Sampling theory is much simpler than systems theory and control theory because it can be explained in simple terminology and proven by mathematics. Understanding sampling theory does not require the same level of experience and broad knowledge as is required for systems theory, and neither does it need such advanced mathematics as is used for control theory analysis (e.g., Laplace transform, Z-transform, Nyquist analysis, state space analysis). Nevertheless, I have not seen any control system specification that adequately covers this critical subject. When I ask my colleagues how they cover the process control sampling requirements in their systems, the following replies are normally given:

- Leave it to the vendor to decide.
- Oil/gas processes are slow, and there is no need to worry about sampling rate.
- I have never had any problem with sampling rate.
- Sampling is not an important issue.

It is common knowledge that some process parameters are slow, and a sampling time of 1–5 seconds is adequate. Examples are levels in large vessels and temperature in big heaters; however, several fast processes need careful analysis in order to select optimum sampling rates for their control. Examples of processes with fast sampling requirements are the following:

- Level in small vessels
- Level in steam drums
- Compressor surge
- pH
- Fast flow
- Some chemical reactions
- Vibration

Such processes may require sampling rates of 4–16 per second. It is tempting to specify a standard fast sampling rate (e.g., 16 per second) for all control loops in order to avoid all of the associated problems. This approach may be feasible in the future, but the present-day control systems are not designed to handle such fast sampling rates. A PCS unit, which is designed to handle 100 control loops, may only accommodate a dozen loops if sampling rates of 8 or 16 per second are used; however, the present advances in electronics and computing (e.g., RISC, 64-bit processors, and modularization) will make the design of ultrafast control units possible. It is only a matter of time before such systems are widely available.

The present-day process-measuring sensors are analog. Consequently, the PCS units have to sample (scan) the sensor outputs at a predetermined rate in order to digitize them and prepare them for further conditioning/processing. The undesirable side effects of digitization are as follows:

- Introduction of nonlinearity
- Reduction of measurement quality
- Possibility of aliasing

Limit cycle is an undesirable effect and is a direct result of nonlinearities. Limit cycle is an output, which persists despite there being no input. It may eventually become periodic and destabilize the control loop. Sampling and digitization of analog process signals create nonlinearities. As a general rule, the higher the sampling time and the shorter the word length (i.e., A/D conversion and memory word length), the larger the system nonlinearities.

The quality of analog parameters is affected by sampling rate, A/D conversion, and word length for storage in memory. Although for some process variables (e.g., temperature, level) the lower quality may not cause a problem, there are a few applications for which low quality is unacceptable (e.g., fiscal metering and algorithmic [repetitive] solutions of control equations).

Aliasing means that when a control system samples signals, it cannot distinguish between the signal and noise. According to sampling theory, aliasing (or folding of frequencies) happens when the difference between signal and noise frequencies is an integer multiple of sampling frequency:

$$f_{Noise} - f_{Signal} = 2n\Pi f_{Signal}$$
$$n = 1, 2, 3, \ldots$$

In order to avoid aliasing, the signal must be filtered by a first-order analog filter with a time constant at least twice the sampling time before it is sampled. For example, if the sampling rate is 2 or 3 per second, a first-order analog filter with a time constant of 1 second is employed.

I would recommend a control system with the following features in order to avoid the sampling/digitization problems (e.g., aliasing, limit cycle):

- Sampling rate of 48 per second
- Antialiasing filter of 50 msecond
- A/D conversion of 16 bits
- Memory storage word length of 32 bits
- Proportional integral derivative (PID) control loop scanning of 1 to 32 per second, depending on process requirement

In order to satisfy the first two requirements, it is necessary for the control unit to employ I/O processing cards that can sample, filter, and characterize process signals independent of the main CPU. In such a control unit, the CPU will handle only the control algorithms, rather than spending its time processing the I/O. With rapid advances in computer hardware and increase in memory size, the second two requirements are no longer difficult to achieve.

3.2.2 Control Theory

It is not my intention to explain control theory in detail here because it is beyond the scope of this book. Besides, numerous books cover the subject at varying degrees of complexity. To fully appreciate control theory, one must have a good understanding of Laplace-transform, Z-transform, modified Z-transform, and state-space analysis. Such analysis provides insight into the control of complex and interactive processes.

The performance of difficult control loops, such as processes with large deadtime, variable gains, or highly interactive elements, can be improved significantly by first modeling them and then analyzing them. Analysis can be carried out by applying readily available software packages. Simulation is normally used to study the effect of changes in the control loop or control parameters. It is also possible to employ dynamic (online) simulation, especially when the control loop has to cope with large and variable nonlinearities.

Control theory in its simplest form relates the process parameter (e.g., flow, pressure, and level) to the PID controller output (C) according to the following equations:

$$C = K(e + 1/T_i \int e \, dt + T_d \, de/dt)$$

$$C = Ke(1 + 1/T_i s + T_d s)$$

$$C_{n+1} = C_n + K[(e_{n+1} - e_n) + T/T_i(e_{n+1}) + T_d/T(e_{n+1} - 2e_n + e_{n-1})]$$

$$s = (1 - Z^{-1})/T$$

$$C = Ke[1 + T/T_i(1/(1 - Z^{-1}) + T_d/T(1 - Z^{-1}))]$$

Where:

e = error = X − Y
PV = Y = Process variable
SP = X = Setpoint = Desired value
K = Controller gain
T_i = Controller integral time
T_d = Controller derivative time
T = Controller sampling (scan) time
s = Laplace-transform operator
Z = Z-transform operator

The previous four equations show PID feedback algorithm in time-domain, Laplace form, digital form, and Z-transform. In analog instruments, operational amplifiers with resistors and capacitors are used to solve the PID equation. In a PCS unit, the third equation is solved by the finite difference method. At every sample time, the controller output (C) is calculated from the present and last sample time error values. Where auto-tuning is employed, the controller parameters (K, T_i, T_d) may be calculated based on some changes in process or operator input. The application engineering aspects of control systems (i.e., advanced control, simulation, knowledge-based systems, and neural networks) are covered in Chapter 6.

In solving the control equation by using sampled data systems, we should pay particular attention to sample time (or controller scan time). The PID controller scan time should meet the process requirements. The first generation of DCS units employed a fixed scan rate. The problem with a fixed scan rate is that if the process is faster than the scan rate, the control performance will be poor and may even destabilize the process. If the scan rate is fast enough to satisfy the fastest control loop, then the control unit may be unduly overloaded. I recommend designing a control unit with an adjustable scan rate of 1–16 Hz and choosing the following strategy:

1. Assign scan rates of 1, 4, 8, 16 per second for processes as indicated below.
2. Use scan rate of 1 for slow processes (i.e., temperature, large vessel level).
3. Use scan rate of 4 for normal processes (e.g., flow, pressure, small vessel level).
4. Use scan rate of 8 for fast processes (e.g., fast flows, level in steam drums).
5. Use scan rate of 16 for very fast processes (e.g., surge control, pH control).

The above scan rates recommendation can be applied when accurate requirements are not known. For example, they can be used at the startup of a plant and improved as the application engineer gains process insight. Scan rates lower than 1 and faster than 16 should be avoided because the former may degrade the control system performance, whereas the latter may overload the control unit unduly.

3.2.3 Information Theory

I suspect that very few, if any, of this book's readers have visited a process plant employing pneumatic control and instrumentation systems. Analog electronic instrumentation (Foxboro Spec 200) was introduced in the late 1960s and DCS (Honeywell TDC 2000) in the mid-1970s.

I remember my first visit to an oil production unit in the Persian oil fields in 1972. All instrumentation and control was pneumatic. The control panels were equipped with large controllers, indicators, and recorders. Hundreds of copper tubes were carrying transmitter outputs from the field to control room panels and from control room controller outputs to field control valves. Large trays were used to neatly bundle the instrument air tubes.

There are numerous restrictions in the application of pneumatic control and instrumentation systems:

- System size
- The number of inputs/outputs
- System modifications
- System expansion
- Space requirements, especially for offshore applications
- Reliability and availability
- Application of advanced control, optimization, simulation, and so on

In the present-day C&I systems, all these restrictions have been tackled by using multiplexing, compact electronic cards, visual display unit (VDU)-based operator stations, flexible alarm systems, LANs, fieldbus, voting, hierarchical design, and software/algorithmic solutions. In an offshore oil/gas production platform where I was the systems engineer during 1986–1988, the control/safety system (PCS, PSD, ESD, F&G) handled more than 30,000 inputs/outputs. It is difficult to imagine what size the central control room (or the platform) would have been if the system had been pneumatic instead of microprocessor-based.

Of course, if we had employed pneumatic C&I, we would have restricted the number of I/O to perhaps 3,000. There are two important issues we have to appreciate here:

1. The application of computers, LANs, fieldbus, smart technology, cathode ray tube (CRT), and LCD-based operator interfaces has facilitated the substantial increase in the number of I/O that C&I systems can handle efficiently.
2. To fully utilize the vast amount of data available in a C&I system, the person–machine interface must be designed carefully.

There is no point in developing a C&I system with a large database (at a substantial cost) and then underutilizing the available information or presenting it poorly to end-users (operators, engineers, and managers). Bombarding operators with a vast amount of data in poorly designed displays is as damaging as providing them with inadequate information. Graphic displays, mimics, alarms, reports, and tabular displays should be developed based on a logical and hierarchical design. In the design of operator interface, we have to consider all cases (i.e., normal operation, startups, shutdowns, emergency shutdowns, panics, and abandonment).

The operator interface requires a teamwork design. The vendor, system designer, and operator must be fully involved in the development of the C&I person–machine interface. The development of graphics, mimics, alarms, reports, and so on could take 6–12 months, depending on the system size and complexity. I recommend that one of the operators (a chief operator or operations supervisor), who has experience with the operation of similar systems, should actively participate in the operator interface design.

The information theory we discuss here is concerned with data, information, presentation, and knowledge. Raw process signals are filtered, characterized, and presented to users in useful and understandable formats. Effective feedback and learning mechanisms should be incorporated into the system. Information and knowledge have been replacing capital, labor, and other physical resources in the last few years and will continue to do so to a larger degree in the future.

The reader should not get confused between what is described above and Shannon Theory.[13] The latter, which is known as communication theory, describes the recovery of information from noisy environments, whereas the former is concerned with data capture and presentation (i.e., the quantity, quality, and usage of information). The main issues are information availability, usability, accuracy, and interchangeability.

Operators spend most of their time in a passive role, intervening directly only when process changes demand them to do so. The value of a display, which is easy to see, navigate, conceptualize, and control, is obvious. Menus and display hierarchy should be configured to facilitate the following requirements:

- Accurate representation of the plant
- Easy conceptualization by operators
- Logical menu/display hierarchy
- Consistent and predictable operation
- Pleasant and easy-to-use interface

Operators learn the system through concepts, analogies, and relationships. The designers of display system should remember this fact and incorporate the aforementioned requirements into the person–machine interface. The use of colors and shapes to enhance the operator's ability to perceive relevant information is well known. Colors are processed by the brain in parallel, whereas shapes are processed in series. Colors can help highlight critical information; however, because operators may find it difficult to interpret more than three to five colors simultaneously, colors should be used sparingly and consistently. The means to help operators avoid entering erroneous commands should be incorporated (e.g., limits for analog parameters and warnings for digital outputs).

3.3 HIERARCHICAL SYSTEM CONFIGURATION

The first generation of control systems using computers was of centralized or direct digital control (DDC) type. In such systems, all process signals were wired to the computer, and the computer would calculate the outputs and send them to the process equipment (e.g., valves, compressors). Such control systems do not have a strong hierarchy, and consequently, have the following deficiencies:

- If the computer fails, control is transferred to analog controllers. During such periods, the system graphics, reports, alarms, and so on are not available.
- Inclusion of analog controllers in addition to the DDC system is a waste of resources.
- Operators get confused if they have to use two different systems for plant operation.
- Wiring and cabling is expensive.

For these reasons, DDC never became popular, and even those who installed such systems had to replace them with the next generation of computer-based control systems (i.e., distributed control system [DCS]). The first DCS, which was introduced by Honeywell (TDC 2000) in the mid-1970s, was based on a strong hierarchical design. Although the present-day control systems are far more powerful and flexible than in the early days of DCS, the hierarchical structures of systems have not changed at all.

Hierarchical design applies to two important aspects of C&I systems, namely system structure and operator interface. The latter subject was described in Section 2.3.4. It will suffice here to emphasise that the operator station displays must be designed based on a strong logical hierarchy. This is necessary not only for efficient control and monitoring of the process but also to avoid costly shutdowns and production, equipment, and human losses.

The C&I system structure or configuration can be divided into four distinct levels of hierarchy: (1) unit control level, (2) LAN-based operator station level, (3) host level, and (4) management computer level. As far as the levels of hierarchy are concerned, no distinction is made between control (PCS) and safety (ESD, F&G) systems. In a true integrated system (see Chapter 4), the same family of hardware and software is used for all applications (e.g., control, safety, mechanical and electrical packages, metering).

Level 1, *unit control*, is wired directly to process/field equipment (e.g., transmitters, valves, switches) and provides the following functions:

- Process interface
- Digitization of process signals
- Characterization of process variables
- Process control (PID, logic, sequence)
- Interface to levels 2 and 3, LAN-based operator stations and host

The reliability of control units (CUs), especially where safety is concerned, is of utmost importance. Redundancy, a well-designed environment, and robust components are needed for Level 1 units. The system should be designed such that the malfunctions of Levels 2 and 3 do not degrade the Level 1 performance. The CU–process interface needs careful design and the assistance of a senior control systems engineer. The present-day field instruments are far more sophisticated and flexible than the previous generation devices (i.e., analog, pneumatic, hydraulic). The selection of field instruments and their interface to the control system should not be assigned to an ordinary instrument engineer. Rather, an engineer who is conversant with systems theory and sampling theory should oversee the activity.

Level 2, LAN-based operator stations, are the main person–machine interface, and their acceptance by the operator is crucial for the success of the C&I system. Because the operator stations heavily depends on the LAN, a well-planned LAN is essential. LAN flexibility, robustness, speed of transmission, and capacity (e.g., the number of devices, the number of branches) play an important role in LAN operation. Operator stations are by far the biggest users of the LAN, especially with regard to real-time data.

Operator stations should provide a user-friendly environment and include engineering (database configuration) and maintenance (online and offline diagnostics) capabilities. All of the operator stations of the system, located in various control/equipment rooms, should provide total and identical operating/monitoring facilities. In a recent offshore project, in which I was responsible for the design of control and safety systems (PCS, ESD, F&G), the system employed nine operator stations (13 VDUs), and had provision for three more operator stations. One of the principal requirements was the capability for operation/monitoring of the total plant from any operator station on any part of the complex (two main platforms, four satellite/support platforms, and three future satellite platforms). Of course, some of the control/operation functions were under password control in each area as required.

Level 3, host, also called server or supervisory computer, holds the C&I global database. This includes process historical data (e.g., averages, trends, totals, events, alarms), application data, high-level/advanced control programs, and management reports. The host should have adequate capacity (memory, processing, and speed) to meet the initial and future requirements. I recommend specifying a host with 50 percent spare capacity.

Supervisory control schemes, plantwide optimization, and simulation packages are normally handled by the host. If the host employs a UNIX operating system, readily available software packages, such as process simulation and optimization, can be acquired from various suppliers for implementation in the host. Simulation is a valuable tool for operator training, control schemes validation, system testing, and process optimization (equipment size and configuration).

Level 4, management computer, which may be located at the company's headquarters, is either a supermini or mainframe computer. Its functions may include production scheduling, planned/preventive maintenance, stock control, accounts, and personnel records.

Not only must each level of hierarchy be designed properly, but the interfaces between them should also be considered carefully. Reliability, capacity, speed of response, and interfacing within the components of each level, between levels, and to foreign systems need thorough analysis and planning. Vendors of control and instrumentation systems should take advantage of the latest developments in electronics, computer hardware and software, communication, and expert systems (neutral networks, fuzzy logic). It is totally unacceptable and irresponsible if vendors offer—or end-users select—control and instrumentation systems, that were designed more than 10 or even 5 years ago.

As we move from higher to lower levels of hierarchy, the reliability and speed of response become more important. The failure of an ESD control unit once every five years may not be acceptable, whereas the failure of the host once a year will not pose a serious problem. Control units should be able to scan process variable and time stamp alarms with a resolution of 1–10 msec, while an update time of 1 second for operator stations and 1 minute for the host may be adequate.

To a lesser extent, hierarchical design also applies to the operator station and control unit configuration. For an operator station, displays and alarm handling should conform to a logical hierarchy. A possible display hierarchy could be overview, area, unit, group, and point displays. Control hierarchy may be optimization, advanced (multivariable) control, and PID control.

3.4 OPEN SYSTEM TECHNOLOGY

In the current production/manufacturing marketplace, the decisive survival requirements are product quality, delivery on time, cost control, and value-added services. In order to meet these requirements and improve continually, increasing use of automation and information technology (IT) is necessary. In a medium-to-large process plant, 10 to 50 packages (mechanical, electrical) will require automation and perhaps dedicated control systems. Traditionally, each vendor provides its own hardware/software for package control. This results in a substantial variety of hardware/software with many serial links to the plant's main control system (PCS).

In a large offshore complex, in which I was responsible for the design of control and safety systems (PCS, ESD, F&G), the client insisted on using independent control subsystems for all packages. This resulted in the use of approximately 100 serial links within the packages and between the packages and the

PCS. Despite our major effort to develop, engineer, and test these interfaces in a controlled and timely manner, we faced numerous problems, which necessitated a large number of my visits and revisits to vendors for clarification, testing, and retesting. Not only were these visits costly because three to four parties were normally involved, but they were also very time consuming.

How can we avoid such interfacing problems? Is it possible to design a plant C&I system without any serial links? The problem can be resolved by using open system technology. We will first see what *open system* means and then indicate the benefits of using such systems. Before doing that, I would like to stress that some process control systems in the marketplace are true open systems, whereas some are partially open.

Various reports[14,15] have been published to describe open systems and the benefits of using such systems. According to systems theory, the system interfacing, both internal and external, has to be designed properly in order for the system to function efficiently. The only way to achieve this goal is to employ open system technology. In order to have a perfect system—or a totally open system—no serial links should be employed. Serial links need gateways for protocol conversion, which makes the interface task complicated.

The formal definition of *open system* by the International Standards Organization (ISO) is based on a seven-layer model: physical, data link, network, transport, session, presentation, and application. Systems that are compliant to the ISO model can be maintained independently and openly distributed to multiple LAN users. Users can procure hardware/software from various sources and feel confident that these systems will interoperate correctly with each other. Open systems do support networks in multisystem, multiuser, and multiapplication installations.

Figure 3-2 shows the basic structure of an open control system. I have used IEEE 802.3 LAN (also called Ethernet, CSMA/CD) because this is the most popular standard, and most system/PLC vendors offer Ethernet. Some systems employ IEEE 802.4 or 802.5 (Token Bus or Ring), but they are not as popular as Ethernet.

The control units that provide control, logic, and process interface can be configured as a single, dual, or triplicated unit to meet the reliability requirements for various applications. They can be programmed to provide for the control of mechanical/electrical packages, such as generators, compressors, pumps, switchboards, or for such applications such as as metering, optimization, simulation, expert systems, and neural networks. If the system user wishes to add a different hardware—provided it conforms to open system standards—it can be connected to the LAN and treated exactly as a control unit.

To highlight the advantages of an open system, consider Figure 3-3, in which an open system configuration (a) and a traditional configuration (b) are shown. In system (a), the plant's main control system (PCS) control units are used for the control of a mechanical package, whereas in system (b), the package vendor's proprietary hardware is employed.

The gateways, which act as translators between the communication proto-

66 | INDUSTRIAL PROCESS CONTROL

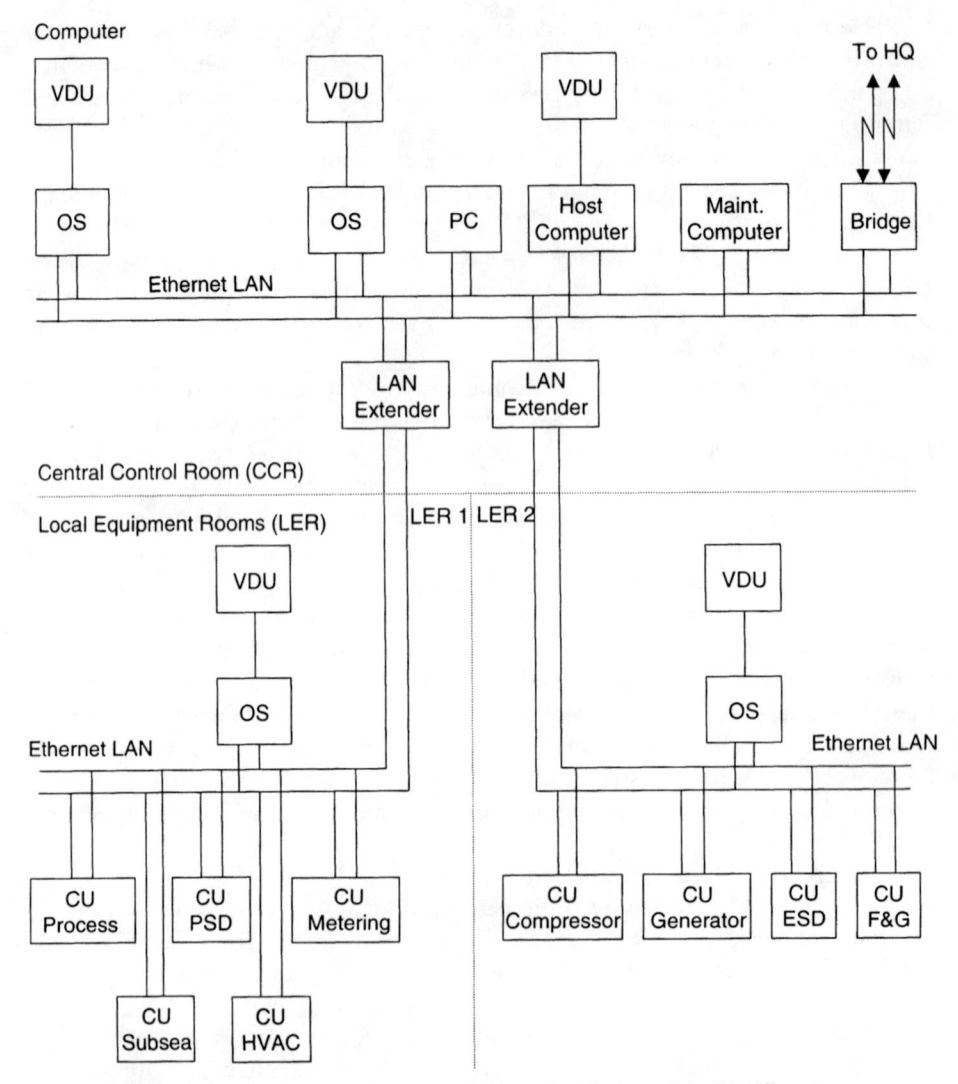

Note: Hardware/software is identical for all CUs; only database (application software) is different.

Figure 3-2 An open control system configuration.

cols of the PCS control unit and the package PLC, are normally single PCB-based units. The different gateway configurations, which I have used in some offshore oil/gas production systems, are as follows:

- Two external gateways
- One external gateway and some internal communication card(s), as per Figure 3-3b
- Two internal sets of communication cards

Systems Theory | 67

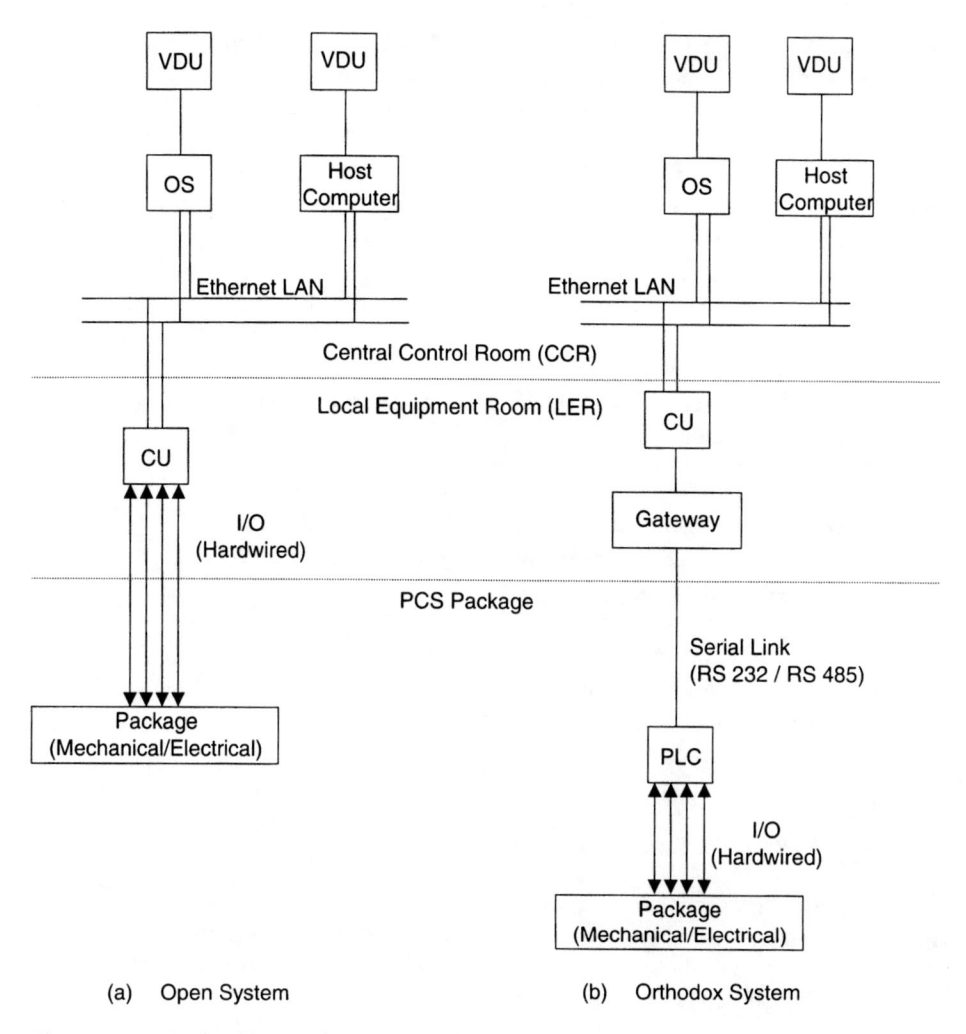

(a) Open System (b) Orthodox System

Figure 3–3 Package control system configuration.

The system with two external gateways is the worst possible configuration, and it must be avoided if at all possible. Because gateways do not provide a complete interface between the package PLC and PCS LAN, an additional PC (or portable computer) is required for engineering (database configuration) and maintenance (diagnostics) of the PLC. Gateways are normally single units; therefore, a gateway failure will cause the loss of interface between the PCS and the package, which may necessitate the package shutdown. Even if a shutdown is not necessary, for some applications (e.g., unstaffed offshore platforms), a prompt service will be required. In such cases, the costs associated with maintenance, transportation, and so on are considerable.

In the case of an open system (see Figure 3–3a), the PCS control unit is employed for control of the package. Although some package vendors resist the use of a third-party controller in lieu of their own system, this can easily be overcome by adding a clause to the request for proposal or specification that the package control shall be handled by the plant PCS. I do not know of any vendor, especially for big packages, that will not sell its system because of adding such a requirement. It must be remembered that in a major package (e.g., a generator set or compressor train), the control system is only a tiny portion of the total system cost, perhaps 5 percent or less. Although vendors will try hard, by introducing various excuses, to include their own control system in the package, it seems illogical to reject a major bid for the sake of a small subsystem.

The use of the main control system (versus package-proprietary PLC) for the control of electrical, mechanical, metering packages has many advantages, including the following:

- Higher reliability because of the absence of gateways
- Easier maintenance/spare control because of reduction in hardware/software variety (no gateways, no third-party monitoring/diagnostics stations)
- Less spare requirement
- Better operability, maintainability, and flexibility
- Easier design and detail engineering
- Easier implementation of modern software packages, such as neural networks and optimization for package control

3.5 UNIVERSAL OPERATING SYSTEM

Every computer, including SAS control units, operator stations, hosts, and PLCs, uses an operating system. The operating system (or executive software) is a software program whose function is the management and overall control of the computer. It performs the following tasks:

- Memory allocations
- Input/output distribution
- Job scheduling
- Interrupt processing

Operating systems are written in machine code, assembly, or high-level languages. High-level languages such as C are widely used in modern systems. Each operating system is designed for some particular tasks. For example, one operating system may be suitable for office-type tasks, whereas another may be more efficient in handling real-time applications.

Although the operating system is a critical aspect of a computer, most users have no knowledge about the functions and importance of the operating system, especially in the case of control and instrumentation engineers (design, application, maintenance) and operators. Popular operating systems such as DOS or

OS/2 are not suitable for control systems. The operating system for a process control system should have the following features:

- Suitable for real-time applications
- Multitasking
- Parallel processing
- Portability
- Suitable for networking
- Popular support and large user base

Ideally, all C&I system units (control unit, operation station, host) should use the same operating system; that is, each unit should not be using a different operating system. The operating system should have the aforementioned features. Such an operating system will be universally acceptable, and the user will be totally free to choose the required software from various sources. At present, the only operating system that meets these requirements is UNIX.

Most SAS vendors have adopted UNIX for their operator stations and host computers. A few also offer UNIX for their control units. In my opinion, all SAS units should use UNIX (or a similar operating system). There seems to be no real challenge to UNIX in the near future. UNIX has a very large user base and is supported by large organizations.

3.6 HARDWARE VARIETY

The hardware variety affects two important aspects of C&I systems: spare parts control/cost and maintenance. Every item in a control system needs spare parts and diagnostic/maintenance tools. The greater the hardware variety is, the more difficult and costly the spare parts management and maintenance task will be. Poor maintenance, of course, interrupts normal operation and may cause the loss of production and quality, damage process equipment, and pollute the environment or even result in personal injury and loss.

How can we reduce the hardware variety in a large C&I system? The answer is simple: Buy all control and safety systems units (OS, CU, PLC) from one vendor. Most C&I, mechanical, and electrical engineers will not agree with this reply. It is arguable that such major packages as generators, compressors, metering, and switchboards are best controlled by their vendors' proprietary control systems. In my opinion, there is no reason to control packages by their vendors' standard control systems. Not only do such systems yield poor operator interfaces, but they also make the tasks of operation, spare parts control, diagnostics, and maintenance difficult. Additionally, control systems offered by package vendors are normally of poor quality and have inferior processing capability.

Many plants, including North Sea platforms, employ the main control system units for the control of major packages. As explained in Section 3.4, it is best to avoid using proprietary control systems for packages altogether. The main control system hardware (i.e., control unit and operator station) will be employed as a common platform for the implementation of control, logic, data acquisition, diagnostics, and management of all major packages. All of the required application software and database and testing of the subsystems will be carried out in close cooperation with package vendors.

When using proprietary control systems for major packages, the maintenance aspect of the package C&I is adversely affected by the following factors:

- Special training is required for the package control system (hardware and software diagnostics).
- Dedicated PCs are needed for diagnostics and troubleshooting.
- Major packages C&I systems cannot be adequately diagnosed from the CCR. This is a major disadvantage if the package is in a normally unstaffed plant.

C&I engineers pay little attention to the maintenance aspects of control systems during detail design, largely because of a lack of experience by design engineers and partly because of a lack of emphasis by the plant owners/users on the criticality of maintenance. As indicated in Section 1.3, maintenance cost is an important factor and requires full attention from the outset of C&I system design. Maintenance cost may have a coefficient of the magnitude of two or three in the objective function.

3.7 SOFTWARE ROBUSTNESS

When we talk of software in C&I systems, we normally mean application software or database. Software written in machine code, assembly, or a high-level language, such as FORTRAN, PASCAL, C, or ADA, is not the subject of discussion here. Although such software programs may be used by control systems for networking, algorithm development, communication protocols, or diagnostics, we are only interested in application software and database configuration, which are normally developed using some of the following techniques:

- Menu-driven displays with data fields to be completed by an instrument engineer
- Detail displays called by point tag number and data fields to be completed by keyboard entry
- Ladder logic, flowcharts, and so on
- Process-oriented programming languages with easy-to-use statements and routines
- Direct translation (by the control unit) from process documents (e.g., P&IDs, cause-and-effects) to application software

The simpler the task of application software development is, the more robust the database will be. An instrument engineer without any programming experience should be able to develop the database with ease. Because using a database is simple and is normally equipped with error-prevention routines (e.g., data type check, help displays, use of default values, range check), application software is inherently robust.

Although database configuration and application software are much simpler than orthodox programming, this does not imply that software quality codes (e.g., ISO 9000) are not applicable to such C&I software. Indeed, all C&I system specifications should adhere to ISO 9000 codes of practice, including development, revisions, modifications, testing, and documentation.

The C&I system database should be thoroughly tested at various stages of development. These include in-house tests, factory acceptance tests, performance tests, site acceptance tests, and commissioning. The system should employ self-tests and the means to prevent errors in data entry or using incorrect parameters. A simple-to-use simulation package should be provided by the C&I system vendor to verify the application software's validity. All of the documents, such as P&IDs, logic diagrams, instrument index, and I/O schedules, which are used for the C&I system database development, must be thoroughly checked for accuracy.

In order to improve the data entry efficiency and reduce the chance of errors in the database, an electronic means of information exchange between the contractor computer and the C&I system should be employed. For example, logic flowcharts may be transferred via Internet from the systems engineer's PC to the PCS host.

4

Integrated Safety and Automation Systems

4.1 INTRODUCTION

In the 1970s and 1980s, supervisory control and data acquisition (SCADA) systems, process management systems (PMS), or distributed control system (DCS) required a host of different hardware and software packages from various vendors. These subsystems were often incompatible, and the only way to interface them was to use hardwiring. This task required thousands of wires and cables with associated terminations, trunkings, trays, marshalling cabinets, and interface input/output (I/O) cards. Engineering, testing, expansion, and modifications of such systems were nightmares.

Advances in computing and communications allowed the use of serial interfaces (e.g., RS 232) between systems supplied by different vendors. Although this capability significantly reduced the burden of wiring, cabling, marshalling, and testing, it still had three major drawbacks:

1. Protocol development and testing of serial interfaces are costly, time consuming, and may cause delays in commissioning of the system.
2. Extra hardware (i.e., gateways and communication cards) is required, which increases costs and space requirements and reduces the system's reliability.
3. Serial interfaces do not allow full data exchange between the two systems. This implies that some aspects of subsystems (e.g., maintenance, database configuration, and monitoring) cannot be efficiently handled by the main control system, in which case extra hardware is required to handle these tasks.

The inadequacy during the early days of computer-based control systems with regard to interfacing and communication with foreign systems and frustration with the testing and data exchange capability of such interfaces led vendors to develop integrated systems. In the following four sections, we study early integrated systems, current integrated systems, future integrated systems, and the benefits of using integrated systems.

4.2 EARLY INTEGRATED SYSTEMS

In order to appreciate the differences between integrated and nonintegrated systems, it is best to illustrate the case with a real system configuration example of each type. The examples are from my own experience and involvement in their design and testing. A medium-to-large offshore oil/gas production complex will require the following systems and major electrical, mechanical packages:

- Distributed control system (DCS)
- Process shutdown system (PSD)
- Depressurization system (DP)
- Emergency shutdown system (ESD)
- Fire and gas system (F&G)
- Vessel control system (VCS)
- Power generation
- Gas compression (export, injection)
- Oil boosting
- Plant/instrument air
- Hydraulic power unit (HPU)
- Water production
- Heating, ventilating, and air conditioning (HVAC)
- Oil/gas metering (fiscal, allocation)
- Multiphase metering
- Subsea production
- Water injection
- Gas injection
- Uninterruptible power supply (UPS)
- Power management
- High-voltage switchboards
- Low-voltage switchboards

In nonintegrated control and instrumentation (C&I) systems, each system/package indicated will normally employ a dedicated proprietary control/monitoring system provided by its vendor. Each system may be equipped with dedicated operation, monitoring, and maintenance facilities from a local control room (LCR) or a central control room (CCR). These subsystems interface with the main process control system (PCS) via either direct hardwiring or serial links. If serial interfaces are used, these require gateways, which are programmed for protocol conversion by either the PCS vendor, package vendor, or both.

Although serial links significantly reduce the burden of wiring, cabling, and marshalling, they still have the following major shortcomings:

- Gateways may need new software
- Gateways require space
- Gateways reduce the overall reliability
- Poor data exchange with PCS

- Poor vendor liaison
- Special maintenance and engineering terminals are required
- Low speed of response
- Inefficient operation and monitoring of packages
- High cost of spare parts storage and management

Design control systems engineers have been aware of these problems since the introduction of microprocessor-based control systems. During the mid-1980s, in a large North Sea production complex, where I was one of the systems engineers, we endeavored to reduce the inadequacies of serial interfaces by directly tackling the aforementioned problems. The system specifications required that all mechanical packages should be controlled via PCS, rather than by vendors' proprietary programmable logic controllers (PLCs). It was also required that the PCS, ESD, and F&G systems should be provided by the same vendor, and the differences in hardware/software between these subsystems had to be minimized. I believe this was the first time that control systems engineers had tried to design an integrated C&I system.

Although the PCS, PSD, ESD, F&G, and most package PLCs were supplied by one vendor, gateways were still needed to interface the PLCs to the PCS LAN; however, most of the PLCs were of the same family. The I/O cards for the total system were identical. In some cases, it was either impractical to use the nominated PLC (e.g., for fiscal metering) or the package vendors were averse to using other PLCs instead of their proprietary ones (e.g., generators).

This project was a great learning school for me. I became familiar with the problems associated with serial links, gateways, control of major packages, and standardization of C&I systems. The supplier of the system (PCS, PSD, ESD, F&G), who is a large manufacturer of C&I systems, also benefited from the project. I believe they were the first company to offer a true integrated control and safety system (in the early 1990s).

In an integrated C&I system, all of the packages are controlled by the main control system (SAS). The SAS control unit (CU) is equipped with all of the necessary hardware, software, and firmware to satisfy the requirements of PCS, PSD, ESD, F&G, generators, compressors, switchboards, subsea system, metering, and so on. Where necessary, a single, dual, or triplicated CU is allocated to the package. For the development of the application software (database) and testing, the CU is shipped to the package vendor for the required period.

In the mid-1980s, when we designed the control and safety system for the aforementioned project, it was not possible to procure an integrated C&I system; however, the system incorporated several important features of integrated systems. We were able to achieve the following:

- One vendor provided most subsystems (PCS, PSD, ESD, F&G, and some mechanical packages PLCs)
- Standardization of operator stations
- Standardization of communication LAN (highway)

- Standardization of I/O cards (for PCS, PSD, ESD, F&G, and some packages PLCs)
- Major reduction in wiring, cabling, and marshalling
- Standardization of serial interface protocols

4.3 CURRENT INTEGRATED SYSTEMS

For a major offshore oil and gas production complex (1991–1994), I evaluated more than a dozen control and safety systems provided by C&I manufacturers of international repute. All of the vendors claimed that their systems employed open system technology and could be configured as integrated control and safety systems; however, my investigation revealed that only two of the systems could offer a true integrated system configuration. (Refer to Chapter 7 for detailed information on some of the evaluated systems.) These two systems are equipped with the necessary control/computational algorithms and programs for efficient control of such packages as generators, compressors (including antisurge), metering, subsea, and vessels (position, ballast, etc.).

Discussions with various vendors revealed that some of them do not understand or appreciate the benefits of integrated systems versus orthodox ones. Some vendors had their own idea of integrated system configuration. Some claimed that integrated systems are unacceptable for safety systems. I have no doubt that most vendors will offer a true integrated control and safety system in the near future. Those that do not will not survive as independent organizations for long.

Chapter 7 describes several control and safety systems offered by some major C&I vendors. The structure and the degree of integration for these systems are studied. Those systems that employ Ethernet (TCP/IP) LANs may be able to interface with other vendors' safety systems without the need for gateways because Ethernet is the most popular communications standard, and most PLC vendors are developing or planning to develop Ethernet LAN for their systems. Other communication standards (e.g., Token Ring [IEEE 802.5]) are not widely used, and I do not know of any PLC vendor who plan to develop them.

4.4 FUTURE INTEGRATED SYSTEMS

A few years ago, it was impossible to design a true integrated control and safety system because of the unavailability of technology and the industry's failure to appreciate the benefits of standardization. The modern C&I systems, especially those used in large offshore oil/gas production platforms, are so flexible and complicated that without adhering to recognized standards, their proper design, efficient operation, and maintenance cannot be guaranteed. The applicable standards may be covering such apparently easy subjects as wiring, painting, protection, noise immunity, or documentation, or complex matters, such as communication, interfacing, reliability analysis, quality assurance/control, software, hardware, database, and graphics.

Numerous organizations, supported by governments, learned societies, and large companies, are dedicated to providing and improving standards. All aspects of C&I systems will benefit from the standardization efforts. Vendors will be forced by competition to adopt the new standards. Designers/engineers will find the design and selection of systems much easier as a result of highly standardized system features. Software, including application software and database, will become totally portable and hardware independent. Hardware will also be so standardized so that users will be able to choose various items of equipment (e.g., operator station, LAN, PLC, control unit, I/O cards, smart transmitters, and valves) from different vendors without worrying about interfacing and compatibility problems.

The future control and safety systems will be true black box–based systems. A black box will be a powerful computer that can be used as a single, dual, or triplicated unit and can be configured (by software codes or pin selection) to perform the following tasks:

- Continuous process control
- Batch process control
- Logic control (PSD, ESD, F&G)
- Operator interface (by adding a VDU)
- Host/supervisory computing
- Server
- Metering (fiscal, allocation, multiphase)

Maintenance will be extremely easy because only one type of hardware is used. Critical units (e.g., operator station, control unit, and PLC) will use redundancy. This feature is especially useful for unstaffed plants where maintenance may not be carried out promptly after a failure. Maintenance can be effectively reduced to the replacement of a black box. Depending on the size of the C&I system, one or more black box units will be stored as spare. The failed black box will be shipped to the vendor for fault analysis and repair.

For process control units (PLC, CU), I/O cards will be required. There will be only one type of card, which can be configured (by software or pin selection) to provide for analog, digital, and pulse I/O functions. I/O cards will have backup (one-to-one or shared) in order to improve reliability and maintainability.

Some engineers, especially the older ones, may argue that uniform hardware is not suitable, because of the lack of diversity, for safety systems. Engineers who use diversity as a tool against integrated systems have no understanding of the systems theory. This is very similar to the "single-loop integrity" argument, which was used in the late 1970s and early 1980s by those vendors who did not offer a DCS, against those who offered one. Today, nobody mentions single-loop integrity, and we have forgotten all about it. The means to improve system reliability and integrity are redundancy, online testing, and proper maintenance, but not diversity or single-loop integrity.

4.5 BENEFITS OF USING INTEGRATED SYSTEMS

Integrated systems are a direct result of the application of the systems theory. As indicated in Chapter 3, the systems theory is mainly concerned with system interfacing. In true integrated systems, the interfacing problems are minimal. The main interfaces in a C&I system are as follows:

- Vendors and subvendors
- Subsystems to the main system (e.g., via serial links)
- Process to various systems
- Operator
- Management
- Engineering (application, maintenance)

In large, nonintegrated C&I systems, more than a dozen major suppliers of control and safety systems will be involved. Some suppliers frequently employ subsuppliers for some portion of their contract. For example, the supplier of a compressor package may subcontract the compressor control system to a third party, and this subsupplier may employ yet another vendor for the antisurge control system. There are always problems, sometimes with substantial financial and delivery implications, with subcontracting and subsupplying, even though design engineers incorporate necessary clauses into the specification of requirements to reduce the burden of subcontracting.

Subsystems, in addition to the problems of subsupplying, require substantial technical effort in the design, development, and testing of their interfaces to the plant main control system. Although the introduction of serial interfaces and their associated standard protocols (e.g., MODBUS) has made the interfacing task much simpler, nevertheless subsystem interfaces are still a troublesome and time-consuming activity in nonintegrated systems.

Interfacing among process sensors, valves, equipment, and the control, safety, and monitoring systems will be much simpler if an integrated C&I system is employed because the vendors of main control systems are familiar with process interfacing requirements. They employ experienced instrument engineers, whereas package vendors are not interested in the process control and instrumentation aspects of their packages. They may employ part-time instrument engineers or a multidisciplinary engineer (i.e., an engineer who will look after package instrumentation, electrical, and mechanical requirements). Although most problems are normally caused by incompatibility of sensors, switches, IS barriers, and PLC I/O cards—and seem trivial to the inexperienced engineer—they could cause shutdowns and production losses if not discovered and rectified during the design phase.

Although the requirements of operators, engineers (application, maintenance), and management are different, because each requires a different type of information, it is safe to assume that all will suffer to varying degrees if a nonintegrated C&I system is used. This is especially the case in systems where unstaffed plants are controlled and operated from a remote CCR. In nonintegrated systems, the interface between subsystems and the main control system is normally via serial links and gateways. Serial links and gateways not only significantly reduce the speed of response (compared with the equivalent integrated system) but also inhibit full data exchange. This results in transmitting only a subset of the database to the main control system and necessitates a local maintenance PC. In unstaffed offshore satellite platforms, the failure of a local control unit or a minor database modification may necessitate an immediate visit to the remote platform. Such visits are expensive and time consuming because the engineer will need to make a helicopter journey and include the company of two to three extra engineers/operators.

In integrated C&I systems, all of the data at various locations (CUs, PLCs) will be available at the CCR operator station and host computer. The operators/engineers will be able to analyze and diagnose any system errors at the CCR and may be able to rectify the system fault and bring it to a safe state. Unlike unintegrated C&I systems, in which there are restrictions in the quantity and quality of data exchange between the subsystems and the main control system, true integrated systems provide complete and seamless data exchange between all the system units (i.e., CU, PLC, OS, host).

5

Project Engineering of Control Systems

5.1 SAFETY AND AUTOMATION SYSTEMS CONFIGURATION

Although most of the control and instrumentation (C&I) systems designed at present are the nonintegrated type, we can safely assume that in the near future most new systems will be the fully integrated type, as end-users become familiar with the benefits of such systems. I have therefore assumed that it is appropriate to base our discussions in this section on integrated C&I systems. Such problematic areas as serial interfaces and subcontracting, which are significant in nonintegrated C&I systems, will not be addressed here because they do not exist in integrated systems. Figure 5–1 shows the control and safety system configuration for an offshore oil and gas production multiplatform complex.

The system covers a production complex, which includes the central control room (CCR), a satellite platform with fiber-optic link to the CCR, a subsea system with umbilical to the satellite platform, and a shore-based monitoring center with fiber-optic link to the CCR; however, a large offshore project may include far more facilities and different communications links, including the following:

- An additional production, wellhead, or accommodation platform, with a bridge between them
- Several satellite platforms, located many kilometers away from the main production platform
- Several subsea wells and associated subsea templates and control pods
- An oil export terminal onshore, or a floating ship with necessary control, monitoring, and communication facilities
- Line-of-sight or radio links instead of, or in addition to, fiber-optic cables

In the next two sections, we will discuss the hardware and software aspects of the C&I system indicated in Figure 5–1.

82 | INDUSTRIAL PROCESS CONTROL

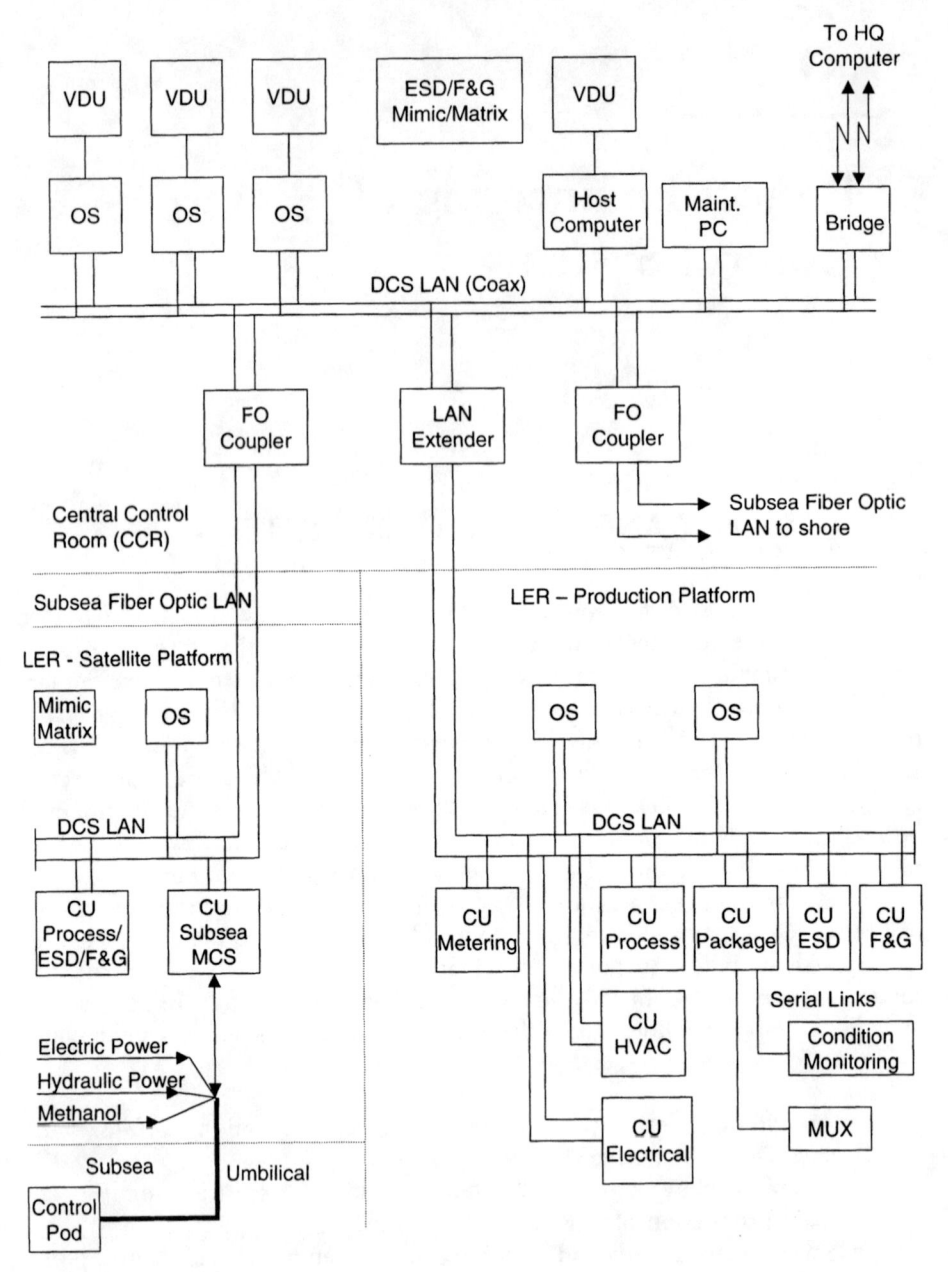

Figure 5–1 Control and safety system configuration for an offshore oil and gas production complex.

5.1.1 System Hardware

The required system components can be broadly divided into five groups:

- Control units (CU), which are used for control (process and mechanical/electrical packages), shutdown (emergency, process, blowdown), and fire and gas systems
- Operator stations (OS), which are the primary operator interface facilities
- Host computer (server, supervisory computer), whose functions include global database management, data archiving, management reporting, and optionally spare parts control and planned/preventive maintenance
- System communications (LAN) and interface to various locations (fiber-optic, line of sight, etc.)
- Engineering and maintenance stations

Each control location will require a different type and quantity of the aforementioned components. There is no universal rule stating how the system components should be distributed. For example, one company may install all control and safety system components for each platform in one control room. Another company may distribute units in several control rooms, where each room will include several CU and OS for the control and monitoring of the associated process, packages, and safety systems. In this case, all local control rooms (LCR) will be linked to the CCR system via a process control system (PCS) local area network (LAN). Although the distribution of system components in several control rooms significantly reduces the cabling effort, in some plants (e.g., processes with a hostile environment or the possibility of releasing toxic gases), housing all equipment in one room may be more desirable. In the following sections, typical control rooms and their components are described.

The CCR is the primary location for the operation, control, and monitoring of the total plant. The CCR is permanently staffed. The design of the CCR is critical, and the comfort of operators should be carefully planned. The ergonomics, console layout, lighting, furniture, colors, and the general layout of the CCR should be of the highest standard. Some firms employ architects and psychologists to help with the design and layout of control rooms and the person–machine interfaces.

The main C&I equipment in a CCR includes the following:

- Operator console
- Control units
- LAN components
- Auxiliary equipment (e.g., printers, recorders, telephones, faxes, power supplies, etc.)

Figure 5–2 shows the console configuration for a typical offshore oil and gas production complex. The size of the console and the number of OS depends on the size of the process and the company's requirements and habits. The

84 | INDUSTRIAL PROCESS CONTROL

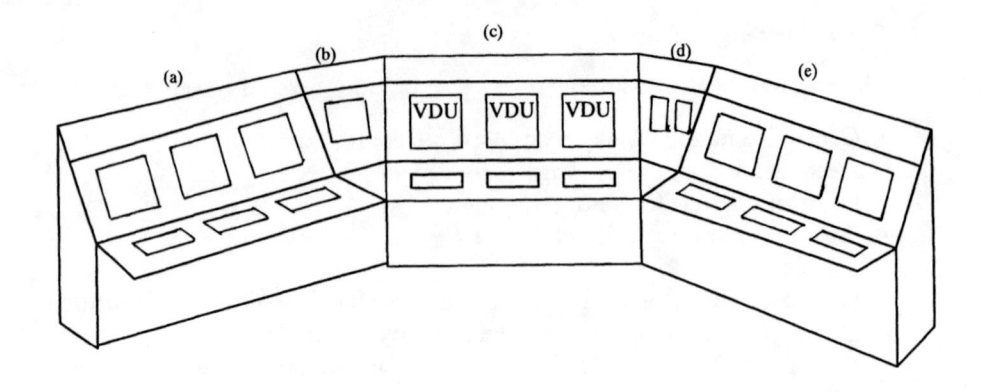

(a) Safety (ESD / F&G)
(b) Telecomms
(c) Process (general)
(d) Telecomms
(e) Utility (generators, switchboards, HVAC)

Figure 5–2   Console configuration in a CCR.

present-day operator stations are powerful enough to configure the plant's total input/output (I/O) and display database in a single station (i.e., any OS in any of the control rooms can be used to monitor and operate any part of the plant). The minimum number of OS in the CCR is two, in order to meet the system availability and operability requirements. Each OS may be connected to one or two visual display units (VDUs) and printers, depending on the number of operators and process requirements.

The console, in addition to OS and VDUs, may also house telecommunications equipment, such as terminals, fax machines, and telephone sets. Each section of the console may be assigned to a particular part of the system (e.g., ESD, F&G, PCS). Alternatively, one or two VDUs may be allocated to each system or a package/process area (e.g., metering, generation, and switchboards).

The CCR may also include system and termination cabinets to provide for marshalling of process I/O and control of some parts of the plant. It is advisable to provide partitioning between the console and the termination/system cabinets. This will not only yield a tidier environment for operators, but it will also eliminate the possibility of disturbance/interference during maintenance. The host computer, printers, and recorders may be located in the CCR or an adjacent room (e.g., the supervisor's room).

LCRs will house control and monitoring equipment needed to control and operate local process and associated equipment and mechanical/electrical packages. The number of LCRs and the type of control/monitoring equipment in LCRs depends on the following factors:

- Type of process
- Size of plant

- Plant configuration
- Process, mechanical, and electrical packages within each area
- Plant owner's preferences and special requirements
- Environmental considerations

Where a major package (e.g., generators, export/injection compressors, switchboards) is covered by an LCR, the possible control and monitoring equipment will include the following:

- System cabinets with several control units (for PCS, ESD, F&G)
- Marshalling cabinets to provide termination for I/O
- An operator console with an operator station, one or two VDUs, a printer, a telephone, etc.

For smaller LCRs, an operator console is not normally required. All of the LCRs and the CCR are equipped with LAN bridges or routers to provide interfacing between all of the C&I system components. Depending on the distances between various LCRs and the CCR, either fiber-optic or line-of-sight/radio will be used for the LAN. Fiber-optic is the preferred means of communication because it can carry more information and at much higher speeds (e.g., 10 MBaud or even higher). It is also less prone to noise interference.

A shore monitoring center (SMC) and export terminal will require an operator console and some control system/marshalling cabinets. Depending on the distance between the CCR and the SMC, either fiber-optic, line-of-sight, or radio will be used to provide an interface between the SMC and the offshore installation.

5.1.2 System Software

The word *software* has a broad meaning and can imply any of the following:

- Operating system (e.g., UNIX, NT, DOS, PS/2)
- A program written in machine code or assembly language
- A program written in a high-level language, such as FORTRAN, BASIC, C, ADA
- A control unit's table of algorithms held in EPROM, known as firmware
- A PCS database (e.g., I/O configuration, graphics, alarm displays, and reports) developed by using the system's configuration tools, known as application software

In the design and engineering of C&I systems, whenever we talk about software, we usually mean database configuration (application software). In modern control systems, the database configuration does not require any software or computer expertise. Personnel who develop the database should preferably be control and instrument engineers. The engineers who use piping and instrument diagrams

(P&IDs), I/O schedules, instrument index, and flow diagrams to develop a database need not have any knowledge of operating systems, machine codes, FORTRAN, ADA, and so on. In most cases, the database configuration is carried out by using menus and fill-in-the-blank fields in the configuration displays.

This discussion does not imply that database configuration is not important or should be handled by inexperienced or unqualified engineers—far from it! Configuration engineers should be chartered engineers who are conversant with P&IDs, instrument index, logic diagrams, and application of the system. The development, revision, and maintenance of the database is not different from any other software system and should be carried out in compliance with ISO 9000/BS 5750 standards.

The vendor of the C&I system should provide a software handling procedure for review and approval by the owner/consultant. Normally, the vendor develops the database and holds the master copy on disk or tape. The database may be revised at site, offshore, or in the vendor's office. I recommend an audit of the C&I system database configuration by an experienced software house. Such an audit will spell out shortcomings on the database configuration and provide long-term benefits.

5.2 CRITICAL PATH METHODS

5.2.1 Introduction

The critical path method (CPM) chart for a C&I system shows all of the project's main milestones with estimated dates for start and completion of each activity. The main activities for the design of a control and safety system are as follows:

- System analysis (initial systems engineering)
- System philosophies
- Specification of requirements
- Potential system vendors
- Bid preparation and evaluation
- Input/output schedule
- Process interface design
- Staff training
- Hardware procurement
- Database configuration
- Application programs
- Tests
- Input/output wiring
- Commissioning

All control systems engineers understand that some activities must be completed before others can start; however, most engineers do not appreciate the impact of revisions of some activities on others. For example, big changes in the I/O schedule may cause substantial revisions to process interface, hardware, and I/O wiring. The systems engineer must foresee changes to process requirements and incorporate adequate provisions and flexibility in the system hardware and software to handle the revisions.

5.2.2 System Analysis

System analysis, or initial systems engineering studies, is a scientific method of study to clearly define the control and safety system's objectives, design methodology, operator interface, staffing, training, control room philosophy, maintenance philosophy, spare parts requirements, test procedures, and so on. Other aspects of C&I systems, such as power supply, earthing, noise, and environmental conditions, must also be considered.

5.2.3 System Philosophies

The present-day control systems are extremely flexible and powerful; their full capabilities are seldom utilized. Most design engineers and end-users are unfamiliar with the system capabilities. Consequently, all of the recently installed C&I systems are underutilized. The development of control philosophies for process unit operations requires a thorough understanding of the process and familiarity with the control system's library of algorithms and various software packages. It is recommended that the task be assigned to an experienced control systems engineer. The engineer must be familiar with the process and conversant with modern control methods and advanced control techniques such as simulation, optimization, and neural networks. He or she should know how to apply these tools in order to improve the operation, maintenance, safety, and production of the plant, and reach the goals of the project.

A powerful tool to study the performance of the C&I systems and to ensure that the project objectives will be met is simulation. The simulation package can be used to evaluate alternative control strategies and to train plant operators.

5.2.4 System Requirements

The control system specification is a fundamental document that should be prepared by a qualified and experienced control systems engineer. Obviously, the engineer must be conversant with systems theory, sampling theory, information theory, and process requirements. The engineer should also be knowledgeable about typical control and safety systems available in the marketplace.

The specification should concisely and unambiguously indicate all of the system requirements. If all of the important requirements are not clearly detailed, the specification will be open to interpretation and may result in the following problems:

- Some vendors, in response to a request for proposal, may undercut other vendors by offering an inadequate system.
- The selected system may require major revisions, even change of major system components or substantial increase in hardware, and costs or delays in startup.
- Serious disagreements may develop between various parties during later stages of the project.

The contents of the specification for a typical control system are indicated as follows:

- A brief description of the process
- The systems objectives
- Approval authorities, regulations, standards, and codes of practice
- Reference to relevant documents and drawings
- System configuration, indicating various locations' control and operation/monitoring facilities
- Interfacing to other systems
- Environmental conditions
- Documentation requirements
- Warranties
- Responsibility for major activities (e.g., tests in factory and site, commissioning, database development, maintenance, spare parts)
- Test requirements
- Input/output schedule
- Expansion/future system requirements
- Marshalling cabinets/process interface
- Operator stations/console/person–machine interface
- Response times for control loops, analog and digital inputs/outputs, displays, alarms, etc.
- Programming requirements
- Optimization packages (e.g., linear programming, hill climbing, etc.)
- Simulation requirements for control, testing, and operator training
- The number and type of displays, reports, alarm summaries, etc.
- Diagnostic and preventive/planned maintenance requirements
- Reliability and availability analysis
- Training requirements for operators, plant engineers, and project/contractor engineers
- Any specialized control requirements (e.g., optimization, simulation, initialization, surge control, flow calculations, choke valve control, postmortem analysis/reports, advanced control such as auto-tuning, neural networks)
- Power supplies/utility

After the vendor is selected and during system development, the specifications will undergo some modifications. In a well-prepared specification, the modifications will be minimal; however, if the specification contains ambiguous statements or an incorrect vendor is selected, it may require substantial revisions.

5.2.5 Applicable Regulations, Standards, and Codes of Practice

This section provides readers with a comprehensive list of standards and codes that are applicable to C&I systems. Some codes (e.g., statutory regulations and certifying requirements) may vary in some countries, whereas others are universally accepted. Most suppliers of C&I systems are familiar with these codes and generally conform to them. When a vendor cannot meet a particular standard, the deviation should be clearly stated and either an alternative acceptable code agreed upon or a concession obtained.

The C&I system specification of requirements should include all regulations, standards, and codes of practice relevant to design, testing, installation, transportation, commissioning, protection, maintenance, operation, safety, environmental issues, and documentation of the system. When a conflict exists between the standards and the specification of requirements, the more stringent requirement should be complied with; however, the following paragraph should be included in order to resolve any possible conflicts:

> In the event of conflict arising between the specification of requirements, the drawings and data sheets, standards and codes, and reference documents, the vendor shall notify the purchaser immediately. The normal order of precedence will be:
> - Attached drawings and data sheets
> - The specification of requirements
> - Statutory regulations
> - Referenced specifications
> - International standards and codes

Statutory Regulations (Offshore Installations)
SI 289: Construction and Survey (1974)
SI 611: Fire Fighting Equipment (1978)
SI 1019: Operational, Safety, Health, and Welfare (1976)
NPD: Acts, Regulations and Provisions for the Petroleum Activities

Guidance Notes (Offshore Installations)
Department of Energy (HSE) Guidance on Design, Construction, and Certification
VMO-DnV Rules for Planning and Execution of Maritime Operations

Certifying Authorities
Lloyds Register of Shipping (Test Requirements for Type Approval of Control and Electrical Equipment)

British Approval Services for Electrical Equipment in Flammable Atmosphere (BASEEFA)
Det norske Veritas (DnV)
Bureau Veritas
Technischer Ubervanchungs Verein (TUV)

Institute of Petroleum (IP)
Model Code of Safe Practice
Area Classification Code of Petroleum Installation

Institute of Electrical Engineers (IEE)
Recommendations for the Electrical and Electronic Equipment of Mobile and Fixed Offshore Installation
Regulations for Electrical Installations

British Standards Institute (BSI)
BS 381C: Specification for Colours
BS 613: Components and Filter Units for Electromagnetic Interference Suppression
BS 1597: Limits and Methods of Measurement of Electromagnetic Interference Generated by Marine Equipment and Installations
BS 3643: ISO Metric Screw Threads
BS 4099: Colours of Indicator Lights, Push Buttons, Annunciators, and Digital Readouts
BS 4683: Electrical Apparatus for Explosive Atmospheres
BS 4800: Specification for Paint Colours for Building Purposes
BS 4937: Industrial Thermocouple Reference Tables
BS 5070: Code of Practice for Engineering Drawings
BS 5260: Code of Practice for Radio Interference Suppression on Marine Installations
BS 5308: Instrument Cables (Parts 1 and 2)
BS 5345: Code of Practice for the Selection, Installation, and Maintenance of Electrical Apparatus for Use in Potentially Explosive Atmospheres (By Part)
BS 5445: Components of Automatic Fire Detection Systems
BS 5501: Electrical Apparatus for Potentially Explosive Atmospheres—General Requirements (Part 1) (EN 50014)
BS 5501: Electrical Apparatus for Potentially Explosive Atmospheres—Enclosure Flameproof "d" (Part 5) (EN 50018)
BS 5501: Electrical Apparatus for Potentially Explosive Atmospheres—Increased Safety "e" (Part 6) (EN 50019)
BS 5501: Electrical Apparatus for Potentially Explosive Atmospheres—Intrinsic Safety "i" (Part 7) (EN 50020)
BS 5501: Electrical Apparatus for Potentially Explosive Atmospheres—Intrinsically Safe Electrical Systems "i" (Part 9) (EN 50039)

BS 5515: Documentation of Computer Based Systems
BS 5536: Recommendation for Preparation of Technical Drawings for Microfilming
BS5555: The International System of Units (SI)
BS 5750: Quality Systems
BS 5760: Guide for Reliability Assessment
BS 5887: Code of Practice for the Testing of Computer Based Systems
BS 6121: Mechanical Cable Glands for Elastomer and Plastic Insulated Cables
BS 6231: Specification for PVC Insulated Cables for Switchgear and Controlgear wiring
BS 6351: Electric Surface Heating
BS 6360: Specification for Conductors in Insulated Cables and Cords
BS 6527: Specification for Limits and Methods of Measurement of Radio Interference Characteristics of IT Equipment (EN 550022)
BS 6667: Electromagnetic Compatibility for Industrial Process Measurement and Control Systems
BS 6883: Elastomer Insulated Cables for Fixing Wiring in Ships
BS 6899: Specifications for Rubber Insulation and Sheath of Electric Cables
BS 60529: Classification of Degrees of Protection Provided by Enclosures
BS 9000: General Requirements for Electronic Components

International Electrotechnical Commission (IEC)
IEC 68-2-27: Test Ea. Shock
IEC 73: Colours of Indicator Lights and Pushbuttons
IEC 85: Thermal Evaluation and Classification of Electrical Equipment
IEC 92-3: Electrical Installations in Ships—Cables
IEC 113: Diagrams, Charts, and Tables
IEC 146: Semiconductor Converters
IEC 157: Low Voltage Switchgear and Control Gear
IEC 185: Current Transformers
IEC 186: Voltage Transformers
IEC 331: Fire-Resisting Characteristics of Electric Cables
IEC 332: Test on Electric Cables under Fire Condition
IEC 337: Control Switches (Low Voltage Switching Devices for Control and Auxiliary Circuits, including Conductor Relays)
IEC 381: Analogue Signals for Process Control Systems
IEC 409: Guides for the Inclusion of Reliability Clauses into Specification for Components (or parts) for Electronic Equipment
IEC 417: Warning Labels
IEC 445: Identification of Apparatus Terminals and General Rules for a Uniform System of Terminal Marking, Using an Alphanumeric Notation

IEC 529: Classification of Degrees of Protection Provided by Enclosures
IEC 605: Equipment Reliability Testing (Parts 1 and 7)
IEC 617: Graphical Symbols for Diagrams
IEC 750: Item Designation in Electrotechnology
IEC 751: Industrial Platinum Resistance Thermometer Sensors
IEC 801: Electromagnetic Compatibility for Industrial Process Measurement and Control Systems
IEC 1508: Functional Safety—Safety-Related Systems

American Petroleum Institution (API)
API RP 14B: Recommended Practice for Design, Installation, and Operation of Subsurface Safety Valve Systems
API RP 14C: Recommended Practice for Analysis, Design, Installation, and Testing of Basic Surface Safety Systems on Offshore Production Platforms
API RP 550: Manual on Installation of Refinery Instruments and Control Systems

Instrument Society of America (ISA)
ISA S5.1: Instrument Symbols and Identifications
ISA S5.2: Binary Logic Diagrams for Process Operations
ISA S5.3: Graphic Symbols for DCS/Shared Display Instrumentation
ISA S5.4: Instrument Loop Diagrams
ISA S50.02: Fieldbus Standard for Industrial Control Systems
ISA S51.1: Instrument Symbols and Terminology

Institute of Electrical and Electronic Engineers (IEEE)
IEEE 802.3: Ethernet
IEEE 802.4: Token Bus
IEEE 802.5: Token Ring

Miscellaneous Standards
MIL-STD-1629A: Procedures for Performing a Failure Mode Effects and Criticality Analysis
OPC: OLE Process Control Standard
ISO 1028: Information Processing—Flowchart Symbols
ISO 7498: Open System Interconnection
ISO 11801: Cabling Systems Performance, Design, and Testing
CCIP/3: Specification for Instrument Panels (Testing Section)

SINTEF STF 75 Standards
F88011: Reliability of Safety Shutdown Systems
F88032: Reliability Evaluation of Safety Configurations
F88033: Reliability of LCC Models for Safety Shutdown Systems
F88034: Reliability Prediction Handbook

F88035: Reliability and Availability of Computer-Based Process Safety Systems
F89023: Reliability Prediction Handbook; Computer-Based Process Safety Systems
F89025: Reliability Data for Computer-Based Process Safety Systems
F90002: Guidelines for Specification and Design of Process Safety Systems

NORSOK Standards
I-CR-001: Field Instruments
I-CR-002: Instrument Systems
I-CR-003: Installation of Electrical, Instruments and Telecommunication
Z-CR-001: Documentation for Operation

5.2.6 Vendor Selection

After the control system's specification of requirements is approved for purchase (or inquiry), ideally a vendor will be selected on an economical basis; however, many factors may affect vendor selection and cause deviation from the ideal case, such as the following:

- The client may favor a particular vendor.
- The client may have prejudice against some vendors.
- The systems engineer responsible for the control system may distort facts during bid evaluation due to favoring a particular vendor or ignorance.

Although it is natural for control systems engineers, or clients, to prefer some vendors to others, it is unfair and irresponsible to deliberately exclude some systems although these systems may be more suitable for the project than the selected one. If the specification truly reflects the project requirements, and more than one vendor can meet the requirements, then the following points should be analyzed carefully before a vendor is selected:

- System cost
- Extra features offered and future expansion capabilities
- Vendor experience and size
- Past experience with the vendor
- Vendor's use of the latest developments in control, computing, and electronics (e.g., open systems, integrated systems, neural networks, fast processors, fast communications)
- Proximity of vendor to the plant and presence in the country

Although these factors are important in vendor selection, items expansion capabilities and experience with the latest developments are by far the most critical subjects. It is impossible for inexperienced control systems engineers to appreciate the significance of such features as open systems and integrated system technology, fast and powerful processors, and communications. Integrated systems reduce the cost of spare parts, maintenance, and interfacing and improve the system response times. Open systems improve system interfacing.

The advantages of fast and powerful processors can be described by the following examples:

- System A control unit employs a slow processor; consequently, it cannot provide scan times and alarm time-stamping better than 1 second, especially if the unit is handling more than 100–200 inputs/outputs.
- System B control unit uses one of the latest central processing units (CPUs) with ample RAM. It therefore can scan process signals and time-stamp alarms with a resolution of 1 msecond. It can provide fast scan rates for many process I/O, even if the control unit is loaded with more than 1,000 inputs/outputs.

Slow scan times in control units may cause two serious problems: (1) they may degrade the process control performance, which if it is not discovered and rectified, may cause production disruption, product quality loss, shutdowns, and equipment wear and tear over the life of the plant; and (2) in conjunction with slow response by operator station, they may cause long update times in graphic and alarm displays. Display update times of 20–30 seconds are normally encountered in slow systems during commissioning and performance testing. In such systems, two courses of action are available: either to upgrade the slow CPUs and increase RAM size or reduce the system load by deleting a significant number of I/O and adding extra hardware (control units, operator stations, etc.). Both options are costly and may cause other unforeseen problems.

5.2.7 Process Interface Design

Although there have been improvements and increased flexibility in wiring, cabling, and marshalling of process I/O to control and safety systems since the introduction of microprocessor-based control systems in the mid-1970s,[3-5] a significant breakthrough will occur when the fieldbus is finally adopted. With fieldbus, it will be possible to transmit several dozen process I/O in a single fiber-optic cable or a pair of twisted wires between field and control rooms. Savings in cabling, junction boxes, cable trays, marshalling cabinets, and electronic I/O cards will be substantial.

Various methods of PCS interfacing are used by different plant designers. Major savings in cabling, installation, termination, testing, and so on are possible if the components of the control and safety system are distributed rather than located in the CCR. When the control and safety units are truly distributed around the plant, they will communicate with the CCR operator stations and host via a dual coaxial cable or fiber-optic LAN. When the fieldbus is combined with distributed control systems, the direct wiring and cabling between process sensors/final control elements and CCR are reduced to a bare minimum.

5.2.8 Operator Interface Design

As stated in Chapter 3, interfacing is the most important aspect of a system, and in a process control system, in my opinion, operator interface is by far the most critical system interface. Deficiencies in most interfaces are normally detected and rectified during design, testing, commissioning, startup, and early operation; however, some deficiencies in operator interface may never be discovered.

Shortcomings in operator displays, graphics, alarm systems, and reports will reduce the operator's efficiency and may cause the following problems:

- Operator fatigue
- Reduction in product quality
- Loss of production
- Shutdowns
- Damage to equipment
- Environmental pollution
- Injury or loss of life

I recommend using the services of one of the plant's senior operators during display development, testing, and commissioning. Alternative methods of data display, data retrieval, operator command, and alarm handling should be thoroughly studied, and the most suitable ones selected.

5.2.9 Staff Training

The training of personnel on control and safety systems should cover four different requirements:

- Initial system design and configuration
- Operation
- Maintenance
- Application engineering

The number of personnel required and the extent of training for each category will depend on the plant size, process type, and the philosophies adopted for maintenance and engineering. For example, the larger the plant and the more complicated the process, the more personnel and more extensive training will be required. It is common knowledge that without adequate training on control, operation, maintenance, instrumentation, and process equipment, the plant resources will be underutilized. Lack of proper training may contribute to shutdowns, production loss, damage to equipment, and even loss of life.

Training is not a short-term affair; it is a permanent activity. Large organizations have dedicated training departments, which are responsible for the proper training of personnel. Investment in training will provide job satisfaction, improved productivity, better quality, and a safe place of work.

After a vendor is chosen to supply the control and safety system, a meeting would be organized to establish training requirements. Senior personnel from the owner, designer, and vendor should participate in the discussions. The engineers who are responsible for database configuration should be trained soon after the vendor is selected. Operators, maintenance staff, and application engineers can be trained before the start of commissioning. These personnel should help with commissioning and site testing.

Some companies employ the vendor for the maintenance and application engineering of the system. By using satellite communications, some vendors can provide online diagnostics and database configuration from their headquarters for systems that are installed hundreds or thousands of miles away.

A simulation system is normally employed to train operators. The simulation system is used to retrain operators and train new operators. The training system can either be built by using a dedicated OS and a CU similar to those used for the plant control system or bought from a company that specializes in building simulation systems.

If the latter option is selected, a simulation system more closely representing the plant's process dynamics can be built. This approach is more effective in training operators or trying new advanced control schemes. If the first option is adopted, the plant's application engineer can easily develop simulation models for various parts of the process. Section 6.6.3 describes how, for example, a tank, a heater, a distillation tower, or a valve, can be modeled and simulated.

5.2.10 Application Software Development

The control system application software, or database, is one of the most critical and time-consuming activities. The number of engineers required for developing the database varies from one to a dozen, or even more, depending on the size of the project. For a system with 20,000 inputs/outputs, 10 database engineers may be required. For a large control system, the database engineers should be supervised by a senior systems engineer. This will help ensure that the task is handled in a professional manner and completed on time.

The database engineers will need P&IDs, process data sheets, and I/O schedules for mechanical/electrical packages. The application software (control, logic, interlocks, etc.) will have to be developed by the database engineers. Logic diagrams and ESD/F&G cause-and-effects are used for this purpose.

An efficient database system is necessary in order to reduce the effects of frequent and/or substantial process data changes. Late changes in some areas of the database (e.g., ESD and F&G cause-and-effects) may invalidate previous tests. The transfer of database from the design/engineering company to the vendor is normally via floppy disks or logic diagrams and reports, but it is possible to exchange the database via electronic mail or Internet.

After the vendor has generated the database/application software and tested it (against the database sent by the contractor's systems engineer), the vendor database report may supersede the contractor's database document. In such a case, late process/logic changes are translated directly into the vendor database report. The original database document then becomes obsolete.

5.2.11 System Testing

The testing of control systems is carried out in several phases. After the system has been assembled, the following tests are normally carried out:

- In-house test
- Factory acceptance test (FAT)
- Heat soak test
- Site acceptance test (SAT)
- Commissioning
- Sustained performance test

Before the system assembly, all of the system electronics and components (e.g., cards, panels, power supplies, monitors, bridges) would have been tested after manufacture. Manufacturers of such components test them thoroughly before delivery to system vendors.

The first system test, which is performed after system assembly, is the in-house test. The purpose of this test is for the vendor to satisfy itself that the system meets all of the specified requirements. The in-house test should be similar to the FAT. The vendor normally prepares an FAT procedure for the contractor/client approval.

In several control and safety system projects, on which I was systems engineer, the in-house tests were either not carried out at all or were performed wholly inadequately. Consequently, the FAT took longer than anticipated because some modifications to hardware and software (database, graphics, logic, etc.) became inevitable.

In present-day projects, where alliances between client, contractor, and vendors are fashionable, the tendency is to leave testing (in-house, FAT, SAT) to the vendor. In my opinion, some vendors are not reliable and may fail to test some critical aspects of the control system adequately. Therefore, it is safer for the FAT to be witnessed by the system engineer. It must be emphasized that it is much more costly to rectify faults during commissioning and startup than during in-house test or FAT. Critical failures (e.g., ESD or F&G malfunctions) may delay startup and interrupt production.

Rectifying faults and retesting during FAT are carried out by the vendor free of charge. But all of the work carried out at the site or the plant by the vendor will be charged to the client, and at a higher rate, especially if the plant is an offshore oil and gas platform.

One of the most problematic areas in control and safety systems is serial interfaces. The serial links have to be tested thoroughly during in-house tests, FAT, SAT, and commissioning. Even if a serial link test has been successful during FAT and SAT, the likelihood of shortcomings occurring during commissioning and startup is high because the FAT environment is different from that of the real plant. Such items of equipment as the control unit, gateway, foreign controller (or PLC), LAN, and OS will be fully loaded during commissioning, whereas in FAT these are partially loaded. A partially loaded controller or gateway will not reveal such problems as low speed of response, loss of data, and effect of noise generated by nearby equipment.

5.3 COMMISSIONING

Successful commissioning of a control system is concerned not only with the application of advanced control philosophies but also with ensuring that the process operators accept the system. The best control scheme is worthless if the operator bypasses it. Every endeavor must be made to convince the operator that the system is an ally in improving the plant operation. Once the operator loses confidence in the control system, it is difficult to regain it, which imposes a heavy responsibility on the commissioning team.

It is assumed that the plant operators will have been trained before the start of the commissioning, although classroom training is no substitute for hands-on experience gained during commissioning. During commissioning, the operators can be engaged in using the keyboard by calling up various displays and manipulating various parameters (i.e., setpoints, outputs, modes of control, tuning constants, alarm settings, database codes). Manipulation and adjustment of these parameters are needed for commissioning and setting control strategies.

Commissioning of control systems consists of two parts: those items that are similar to analog instrumentation systems and those that are concerned with advanced control (i.e., optimization, simulation, automatic startup and shutdown of unit operations, decoupling, heat balancing).

The first part must be commissioned and put into operation before the items of the second part can be started. All the I/O will be tested for continuity (from sensor/final control elements to junction box, termination panel, I/O cards, loop display, group display, alarm displays, overview, mimics, etc.). Then all control loops, including slaves in the case of cascade loops, will be tuned; the master loops of cascade control schemes will then be tuned. All the status, alarm, regular, and management reports and logs will be commissioned, too.

At this stage, it is good practice to allow plant operation to continue for awhile, before the advanced control strategies are commissioned. This approach is useful in that the operators will need to pay more attention to, and be more actively involved in, control and monitoring of the plant. When the operators are satisfied with the first stage, the advanced control strategies can be commissioned one by one. It is of paramount importance to explain each strategy to operators and make sure they understand the purpose of the strategy—what is expected to be gained by its application and what must be done if it fails or is bypassed.

As the application engineer learns more about the behavior of the process, he or she can refine the models used for various advanced control schemes. Modeling is a difficult and important aspect of the design of advanced control strategies. These models mimic some parts of process equipment. As an example, dead time simulation for the Smith predictor scheme mimics process dead time. This may be applied to a composition control system or temperature control, where long pipe runs are involved. The more accurate the process dead time estimation, the better the performance of the Smith predictor scheme.

5.4 PROJECT STAFFING

For a control system to be successful and help to achieve the objectives as defined in various project philosophy documents, the correct type and number of staff should be assigned to all system activities, which include the following:

- Design
- System configuration and development
- Database configuration and development
- Testing
- Commissioning
- Operation
- Application engineering
- Maintenance

The number of engineers for each task depends on the size and type of the project (plant); however, the type, qualifications, and experience of these staff members are not affected by the project size. In the following paragraphs, the staff requirements for each activity are explained. The plant owner should ensure that the correct type and number of personnel are assigned to the system during various phases and for all tasks.

For the design phase, a qualified and experienced senior control systems engineer should have overall responsibility. Depending on the size of the system, one or more engineers, who report to the senior control systems engineer, will be assigned to various design tasks, such as PCS, ESD, F&G, database, control/logic diagrams, and serial interfaces. The senior engineer should be conversant with systems theory and have preferably designed a few similar systems. Each of the team members should have experience in the subsystems (e.g., PCS, ESD) for which they are responsible.

The system configuration and development is primarily a vendor activity; however, the senior control systems engineer should have close cooperation with the vendor and ensure that the system configuration is robust. The vendor will assign a project manager, a lead engineer, and several systems engineers for various system development functions (e.g., database, graphics/displays, serial interfaces, testing, marshalling/termination panels/cabling). It is useful to assign some of these systems engineers to site testing and commissioning of the control and safety system.

The database configuration and development can be carried out either by the system vendor (e.g., in alliance projects) or partially by the engineering firm and partially by the vendor. In both cases, it is necessary for the vendor lead engineer and contractor senior control systems engineer to fully monitor the database configuration and development. The integrity and efficiency of control and safety systems largely depends on the quality and robustness of the database. Quality control routines, which are necessary for software development, are also important for database engineering. Documentation, revision handling, reviewing, auditing, and archiving aspects of database development should be designed properly.

Testing and commissioning activities require involvement and close cooperation among various parties (i.e., client, contractor, vendor engineers, and plant operators). It is strongly recommended to use plant operators during testing and commissioning. This will not only help these activities but will also be beneficial to operators. The operation of a process plant depends on two factors: (1) the quality of operators, and (2) the robustness, integrity, and efficiency of the C&I system.

The operators are the primary users of the C&I system. It is vital that they understand the C&I system and the process very well. Operators receive in-house training in various aspects of the process, especially in petrochemical plants, including unit operations and control aspects of plant items. I believe some companies in the United States assign the operation of some plants to young process engineers. Process engineers can learn the process and the C&I system quickly and become efficient operators.

Operators need proper training in the use of the control and safety system. Training can be provided either by the system vendor in its training center or in the plant by the use of simulation. It is a good practice to install a permanent training/simulation system for training and retraining operators and other personnel. It is highly desirable to involve operators in site testing and commissioning of the C&I system.

The application engineer's job is to apply advanced control techniques, such as multivariable analysis, optimization, neural networks, and simulation, to improve the control and operation of the plant. The application engineer should have a good understanding of the process and various advanced control schemes available. Application engineers are essential for plants that have difficult processes that cannot be controlled adequately with simple control methods.

Plants such as petrochemical, cement, pulp and paper, and pharmaceutical, where large dead times, large and variable gains, and interactive parameters make process control difficult, require the services of an experienced application engineer. Application engineers should work closely with the plant operators in order to discover inefficient control loops and try advanced control schemes. Various advanced control packages are readily available to enable application engineers to meet their plant's control requirements. The following list shows some of these packages:

- Optimization (linear and nonlinear)
- Simulation
- Statistical analysis and control
- Fuzzy logic

- Neural networks
- Multivariable analysis

The maintenance engineers and technicians of C&I systems are normally trained by the vendors. The maintenance of control systems is an easy task because the modern systems are equipped with comprehensive diagnostic routines and helplines directly linked to the vendor's headquarters. Most vendors provide 24-hour troubleshooting assistance. The diagnostics will show which electronic components have failed and where the root-cause of the problem lies.

5.5 DOCUMENTATION

The C&I documentation required to design and develop the system can be divided into two groups: those provided by the purchaser/engineering firm and those provided by the vendor. The main documentation prepared by the purchaser or the consultant/engineering company is the following:

- Specification of requirements
- System block diagrams
- Process flow diagrams (PFDs)
- Piping and instrument diagrams (P&IDs)
- Purchase order (PO)
- Logic diagrams
- Hookups
- Loop diagrams
- Instrument index

Documents produced by the vendor are in accordance with the suppliers document register list (SDRL). The SDRL indicates the category of documents and the required dates for vendor's documents. The main documents provided by the vendor are the following:

- Control rooms equipment layouts
- Assembly drawings
- Wiring/cabling/interconnection diagrams
- Wiring schedules
- Functional design specification (FDS)
- FAT/SAT procedures
- Reliability analysis
- Database/displays/reports
- Operation/maintenance/certificate manuals

Some of the vendor's documents cannot be completed before some of the purchaser's documents are completed, and vice versa. For example, hookups and loop diagrams cannot be completed before wiring schedules are ready, and the wiring schedules cannot be completed before the hookups are prepared. Unnecessary arguments often arise because of this dependency problem. Vendors and purchasers/contractors should appreciate this situation and endeavor to resolve the problem with cooperation and early warning of possible revisions. No benefit is gained by blaming each other for the lack of information and progress.

Although the SDRL clearly shows the required dates for vendor documentation, some target dates cannot be met because of dependency on the purchaser's documents; however, such target dates are normally not defined for the purchaser's documents. I strongly recommend that such target dates be agreed on for the purchaser's documents. They should be realistic and adhered to.

In a recent project, in which I was responsible for the design of the control and safety system, although such dates were established at an early stage, we failed to meet most of the target dates and, in some cases, with large delays. This problem not only created frustration and delay in the system testing and delivery but also increased the system cost substantially. Delays in the delivery of the purchaser's documents could result from some of the following factors:

- Unrealistic target dates
- Poor management (e.g., lack of cooperation between disciplines or engineers)
- Inexperienced/unqualified engineers
- Inadequate staffing
- Lack of cooperation by the vendor

Some documents (vendor's and purchaser's) are needed at an early stage in order to complete some plant tasks. For example, hookups and termination schedules are needed to carry out cabling and termination. Consequently, all parties involved (i.e., design, construction, expediting, and project engineer) pursue the matter and ensure that the documents are delivered on time; however, some other documents, which may seem unimportant to the inexperienced, are forgotten and may be produced too late to be useful. Examples of these latter documents are block diagrams, reliability analysis, and test procedures.

Block diagrams are necessary to ensure that all of the system components and interfaces are in place and not forgotten. They also help with system analysis, where engineers from various disciplines/organizations may evaluate the system hardware and software. The block diagrams should be the responsibility of the control systems engineer and should be updated frequently to reflect the latest system changes.

The reliability analysis document should be produced as soon as the purchaser and the vendor have agreed on the system configuration (hardware, software, and interfaces). The document should show in detail the reliability calculations for all of the C&I components, units, subsystems, and the system. It should clearly show the weak points (e.g., common mode failure possibility). Where weak points are discovered, it will be easy to eliminate them by either redesign or using redundancy. This solution may be too difficult at later stages of the project (e.g., during commissioning and operation). I suggest that the reliability analysis document be prepared with the bid and updated after every major system configuration change.

The purchaser will normally have prepared several general documents to be used by all vendors. These documents should be sent to vendors at an early stage, possibly with the request for proposal or after contract award. Examples of these documents are as follows:

- Cabling, wiring, and trunking specification
- Equipment paint, materials, and construction specification
- Packing and shipping instructions
- Environmental conditions
- Document handling procedure
- Equipment numbering

6

Application Engineering of Control Systems

6.1 INTRODUCTION

The distributed control system (DCS) of the 1970s and 1980s provided a useful library of algorithms and software linking of function blocks (FB). This allowed the implementation of simple to advanced control schemes by configuration, rather than by hardwiring or software programming. An instrument engineer could learn, through two to three weeks training, how to configure such control loops as proportional integral derivative (PID), cascade, override, feedforward, heat balancing, dead time compensation, logic, and batch control.

Although simulation, mathematical modeling, and optimization are useful tools and are widely used for such applications as training, analysis of control schemes, auto-tuning, and advanced control, the progress in their application to control engineering has reached a stalemate. Some scientists during the last two to three decades dreamt that the use of ultra-powerful computers to implement simulation and mathematical models would solve difficult process control problems. Processes with variable dead time or gain, for example, can be mathematically modeled and analyzed/simulated to find the optimum control parameters. Because process dead time and gain are nonlinear, however, mathematical models/analysis cannot solve such problems effectively, and the control parameters found by the computer are normally inaccurate.

The process dead time and gain are not predictable because they vary according to the process conditions. For instance, changes in flow, pressure, temperature, or level may cause large variations in process dead time and gain. This makes some process control loops extremely nonlinear. All mathematical and simulation models we use to analyze control schemes assume that processes are linear or near linear.

One of the most widely studied control problems is the tuning of PID control loops. Control engineers are familiar with the analytical methods, such as Bode and Nyquist, of finding optimum controller proportional band, integral, and derivative times. Although the Bode and Nyquist methods are useful in tuning control loops, some control loops do not function satisfactorily if tuned with ordinary methods. If the controller parameters are not configured correctly, the difficult loops are highly likely to oscillate and destabilize the process and cause shutdowns.

The difficult control schemes require constant attention by operators and application engineers. In some cases, operators may transfer the controller to manual mode and adjust its output manually based on their own judgment. In other cases, operators may set the controller setpoint to a value different from the optimum target in order to reduce the possibility of hunting. Readers should appreciate that operators are the key people in managing unstable control loops. Their experience is invaluable and is the basis for the implementation of modern techniques, which use artificial intelligence, neural networks, fuzzy logic, and so on.

In the next five sections, we will first study some typical and advanced control schemes, which are widely used in various process plants. Then, traditional control algorithms and modern techniques such as neural networks and fuzzy logic are explained. Finally, simulation and its application in process control and training of operators is discussed.

6.2 TYPICAL CONTROL SCHEMES

It is not my intention to present readers with many control schemes here because they can be found in various readily available publications. Four popular control loops will be studied in this section. They will then be used to show how we can improve their performance by introducing advanced control schemes in the following section. Figures 6–1 through 6–4 show control systems for level, temperature, pH, and blending, respectively.

Level control is used in all process plants; some are very simple (e.g., a knockout drum or flow tank), whereas some are difficult and have scope for improvement (e.g., multiphase separators). Temperature control is also widely used, especially in chemical plants. In some processes (e.g., chemical reaction and distillation towers), temperature may be critical and any improvement in its control is highly desirable. Although pH and blending control are not as widely used as level, flow, pressure, and temperature control, I have included them here because they are interesting subjects and their control is normally difficult.

6.3 ADVANCED CONTROL

In this section we explain the difficulties associated with the four control schemes of Section 6.2 and show how their control can be improved using the standard algorithms offered by the process control system (PCS) control unit.

6.3.1 Level Control by PID Gap

For large tanks, or similar processes where a small deviation will be rectified by the self-regulatory feature of the system, a PID gap algorithm proves useful. In such a control loop, the controller output does not fluctuate for predefined deviations of process variable from the desired setpoint.

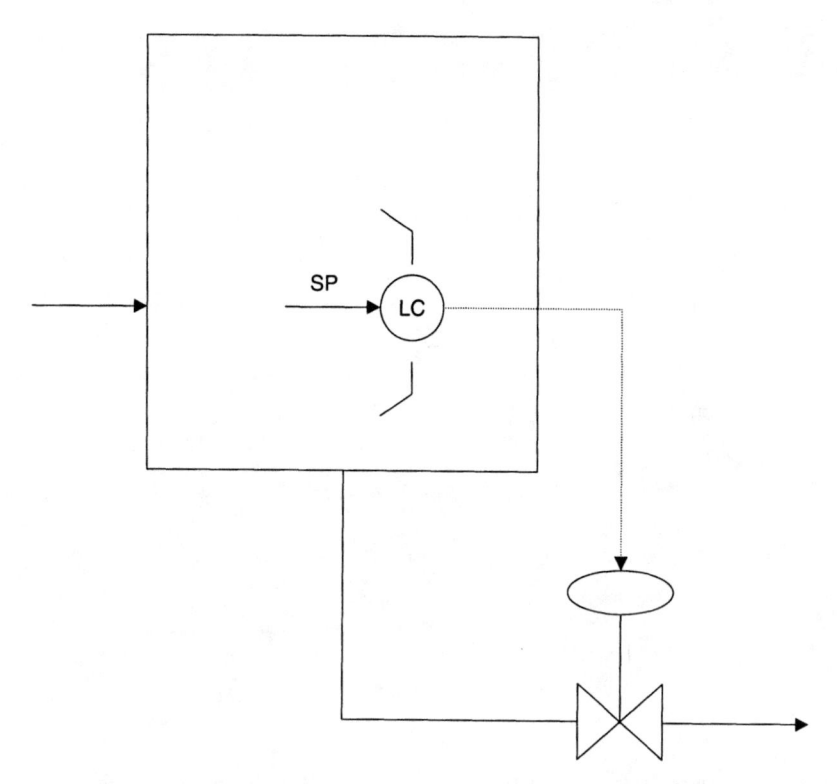

Figure 6-1   Level control system.

The level controller (LC) in Figure 6-1 uses a PID gap algorithm. Controller output (C) is given by the following equation:

$$C = K_g(e + 1/T_i \int e\, dt + T_d\, de/dt)$$
$$K_g = 0 \text{ if } L_2 < e < L_1$$
$$K_g = K \text{ if } L_1 < e < L_2$$

where K = gain
T_i = integral time
T_d = derivative time
e = error
L_1 and L_2 are upper and lower acceptable deviation limits (e.g., +10% and −5%)

6.3.2 Dynamic Feedforward Control

Most process plants employ boilers and furnaces. Dynamic feedforward control schemes can improve the control performance substantially. Figure 6-5

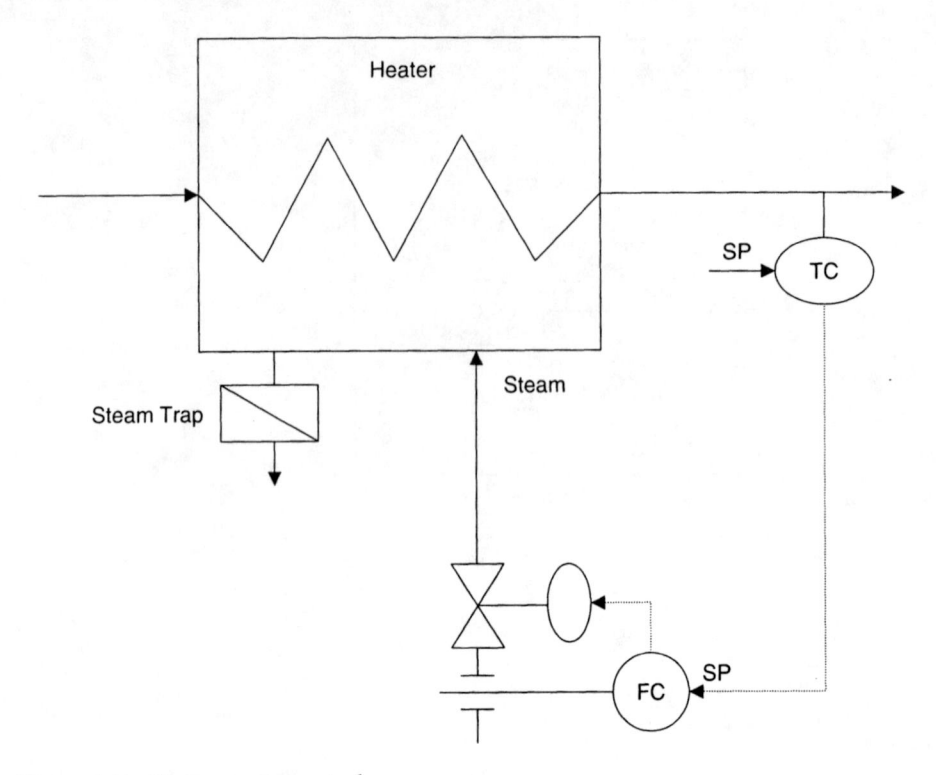

Figure 6–2 Temperature control system.

shows the dynamic feedforward control scheme for a typical heater. The system employs the dynamic element XY-1, which is a lead-lag compensator, given by:

$$\frac{1+T_1 s}{1+T_2 s}$$

Although the dynamic feedforward control system needs more parameters than the simple control loop and seems complicated, in practice all of the required parameters are normally available in the PCS control unit. The plant PCS control unit can be easily reconfigured to provide the desired control scheme.

6.3.3 Heat Balancing Control

In order to improve the efficiency of a furnace, the control scheme shown in Figure 6–6 can be used. The figure indicates a two-pass heat balancing system; however, it can easily be expanded for heaters with more passes. The use of the heat balancing scheme suggested here will not only reduce fuel consumption but will also reduce the possibility of coke development in the heater tubes.

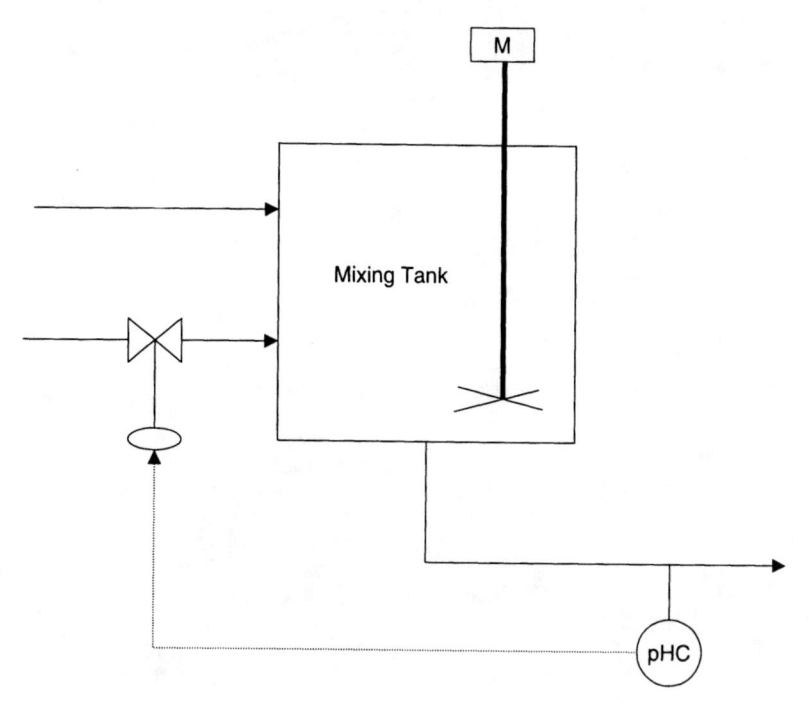

Figure 6-3　pH control system.

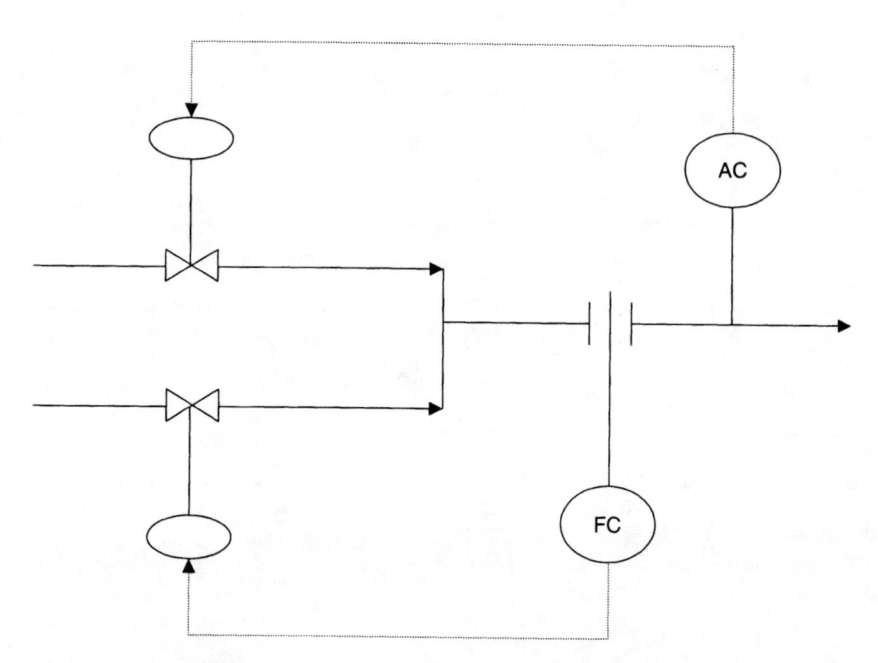

Figure 6-4　Blending control system.

110 | INDUSTRIAL PROCESS CONTROL

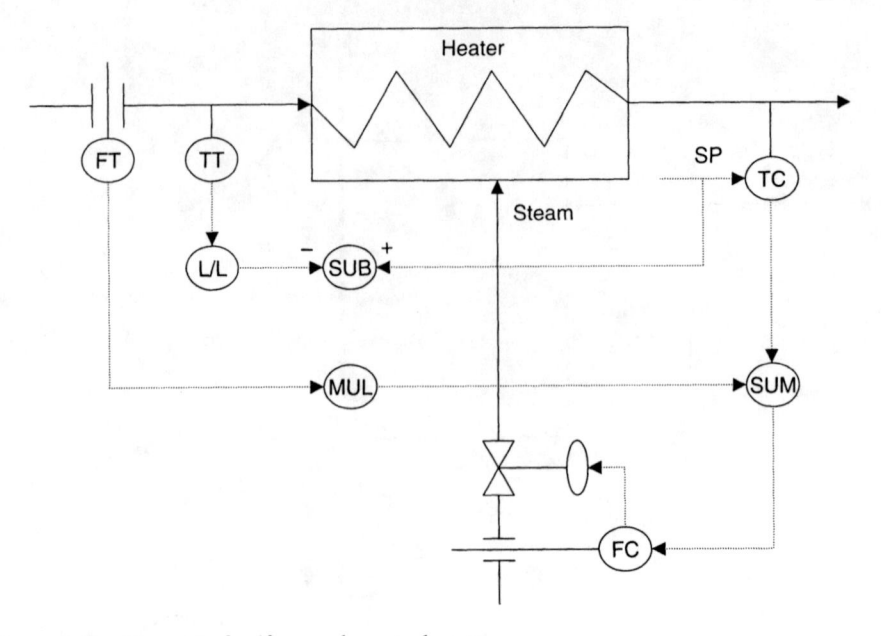

Figure 6-5   Dynamic feedforward control system.

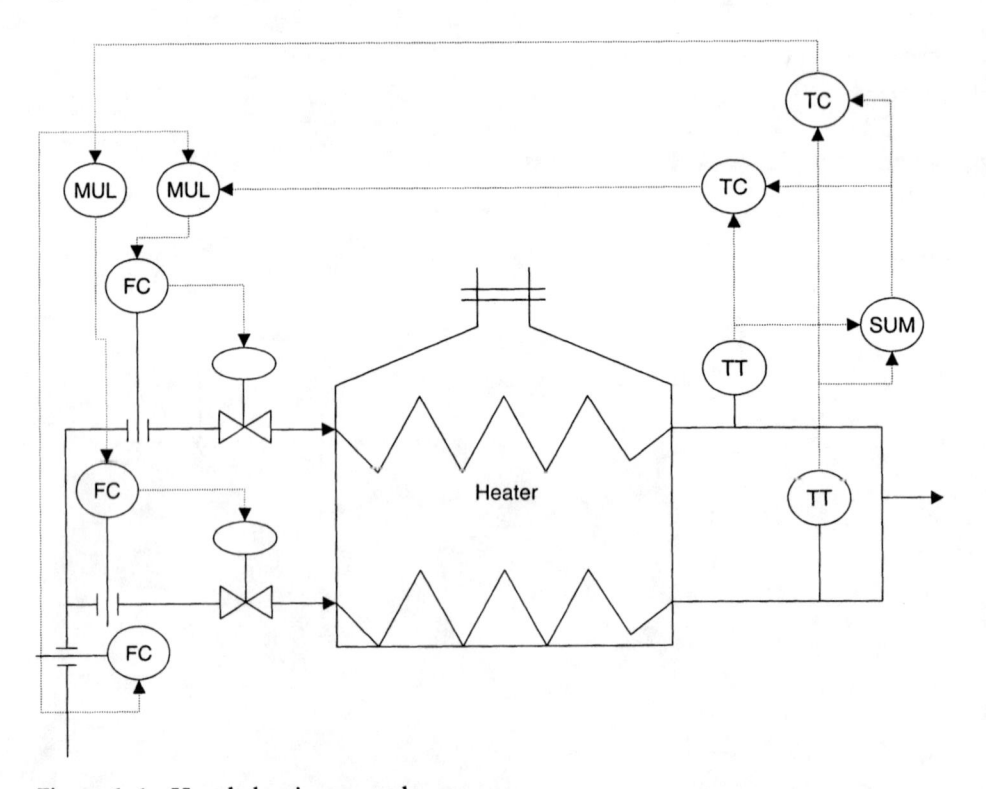

Figure 6-6   Heat balancing control system.

```
      pH
       ^
       |
       |  ──╮
       |    ╲
       |     ╲
     7 |······╲·············
       |       ╲
       |        ╲___
       |            ╲____
       |_____> Flow
```

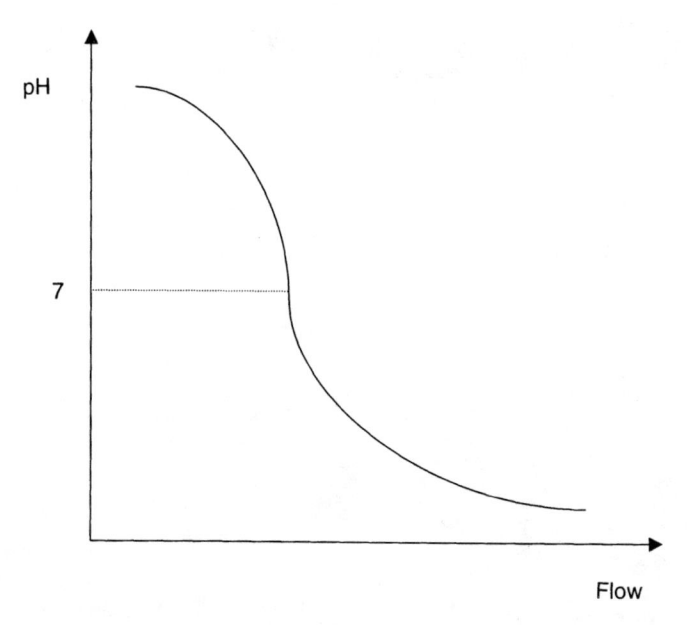

Figure 6-7 pH titration curve.

6.3.4 pH Control by PID Error Squared

pH control is one of the most difficult control systems because of the highly nonlinear feature of the process. Figure 6-7 shows a pH titration curve. The process gain is much higher at pH = 7 than where pH is near 10 or 4. Simple PID controllers cannot control pH efficiently. In order to counteract the variable pH gain, a PID error squared algorithm is recommended. The following equation indicates the PID error squared algorithm:

$$C = K_s \left(e + 1/T_i \int e \, dt + T_d \, de/dt \right)$$

$$K_s = |e|$$

$|e|$ = absolute value of error

This PID algorithm is efficient where the controller setpoint is near 7. At around pH = 7, the controller gain (K_s) is small, where the process gain (K_p) is high. Figure 6-8 indicates the changes in process and controller gains versus pH changes.

6.3.5 Multivariable Control

Multivariable control can be used for processes with interactive variables. A multivariable control algorithm endeavors to decouple interactions between process variables. Relative gain (K_{ij}) is used to measure the degree of interaction.

112 | INDUSTRIAL PROCESS CONTROL

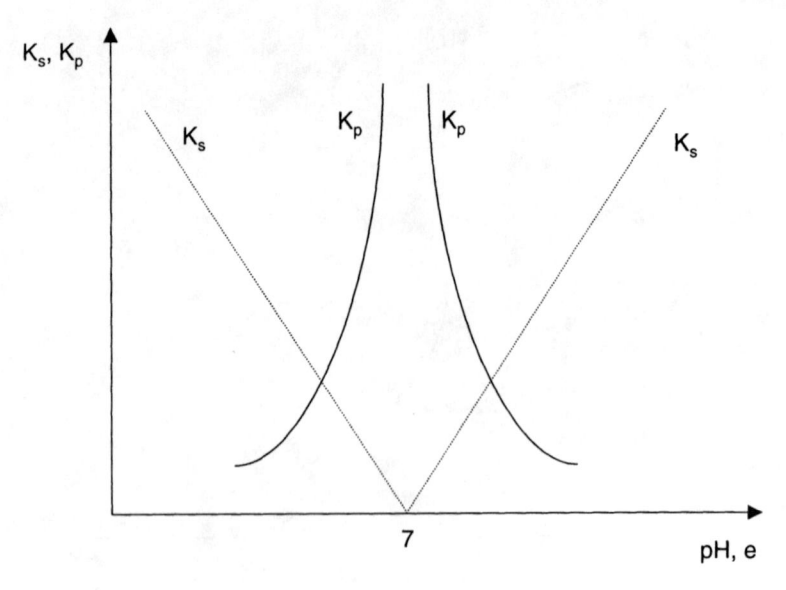

Figure 6-8 Variations in controller/process gain versus pH changes.

Relative gain for a system with Y_i controlled variable and X_j manipulated variable is given by:

$$K_{ij} = \frac{\left.\frac{dY_i}{dX_j}\right|_x}{\left.\frac{dY_i}{dX_j}\right|_y}$$

Where $dY_i/dX_j|_x$ is the sensitivity of controlled variable Y_i to manipulated variable X_j when all other manipulated variables are constant; $dY_i/dX_j|_Y$ is the sensitivity of controlled variable Y_i to manipulated variable X_j when all other controlled variables are constant:

$$Y = (Y_1, Y_2, \ldots Y_m); X = (X_1, X_2, \ldots X_n)$$

The relative gain matrix is given by:

	$X_1\ X_2 \ldots X_j \ldots X_n$
Y_1	$K_{11}\ K_{12} \ldots K_{1j} \ldots K_{1n}$
Y_2	$K_{21}\ K_{22} \ldots K_{2j} \ldots K_{2n}$
$\vdots$	
Y_i	$K_{i1}\ K_{i2} \ldots K_{ij} \ldots K_{in}$
$\vdots$	
Y_m	$K_{m1}\ K_{m2} \ldots K_{mj} \ldots K_{mn}$

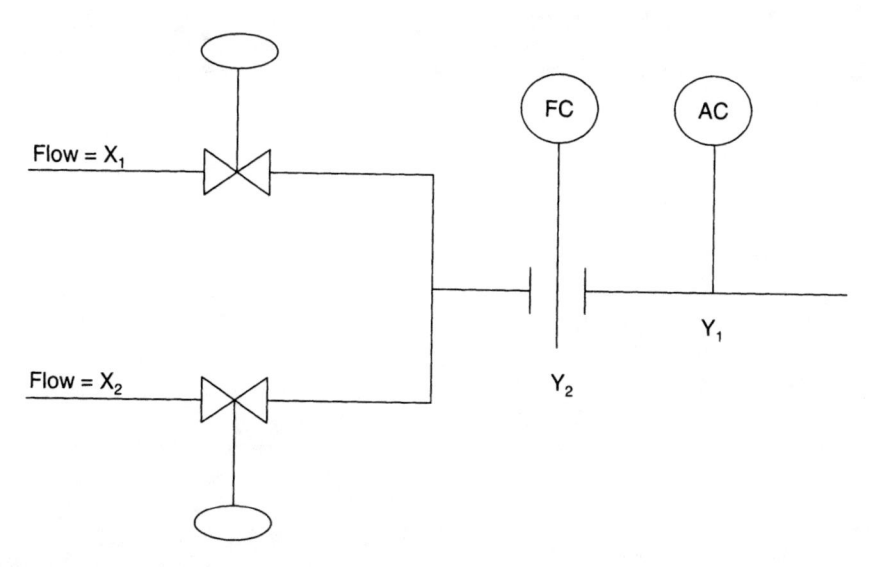

Figure 6–9 Blending process diagram.

The sum of every row or column is one:

$$\sum_{j=1}^{n} K_{ij} = 1 \text{ for } i = 1, 2, 3, \ldots m$$

$$\sum_{i=1}^{m} K_{ij} = 1 \text{ for } j = 1, 2, 3, \ldots n$$

Interaction is negligible if $K_{ij} = 0$ and is high if $K_{ij} = \pm \infty$. For a binary system, a relative gain of 0.5 also indicates a strong interaction.

To show how this model can be applied to a real-life problem, the following blending system will be used (see Figure 6–9). The model is:

$$Y_2 = X_1 + X_2$$

$$Y_1 = \frac{X_1}{Y_2} = \frac{X_1}{(X_1 + X_2)} = \frac{Y_2 - X_2}{Y_2}$$

The relative gain matrix is:

	X_1	X_2
Y_1	$1 - Y_1$	Y_1
Y_2	Y_1	$1 - Y_1$

Figure 6–10 shows the decoupled control system.

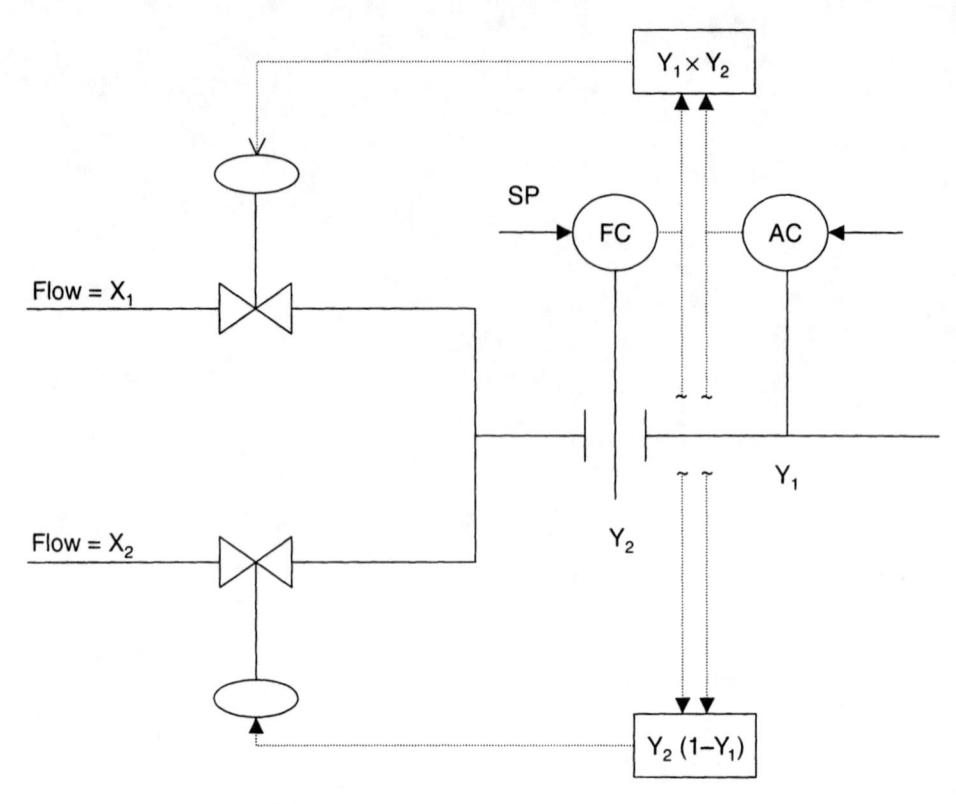

Figure 6–10 Decoupled blending control system.

6.4 TRADITIONAL CONTROL ALGORITHMS

The present-day control systems are equipped with a comprehensive library of algorithms that will satisfy the requirements of most processes. These algorithms reside in the ROM of control units (CU), PLCs, and the host. Figure 6–11 shows such a list of algorithms. Additionally, process-orientated programming languages are also provided, which can be used to implement various control/logic schemes. The control schemes are developed in CU/PLC/host by the use of function blocks (FB). Some vendors may employ different terminology (e.g., computation slot, logic block).

An FB has several inputs and outputs and uses one or more algorithms to satisfy the requirements of a particular control loop. For large control loops, it may be necessary to use several FBs. The inputs of FBs can be from the process (e.g., flow transmitters, level transmitters, manual switches), from the keyboard (e.g., parameters such as PID constants, ranges, alarm settings, manual commands), or from the outputs of other FBs. The outputs of FBs can be used to control the process (valve open/close, motor shutdown), as inputs of other FBs, for indication, warning, alarms, reporting, and archiving.

Available Algorithms	Supported Functions
PID	Modes (Manual, Auto, Cascade, Backup Cascade)
PID with Feedforward	Mode Attribute (Operator, Program)
PID with External Reset Feedback	Normal Mode
PID with Position Proportional	Remote Cascade, Remote Request, and Remote Configurable Per Slot
Position Proportional	
Ratio Control	Initialization
Ramp Soak	Windup Protection
Auto/Manual Station	Fixed or Auto Ratio and Bias
Incremental Summer	Override Propagation
Switch	External Mode Switching
Override Selector	Safety Shutdown
	Target Value Processing (Setpoint Ramping)
	Alarms
	Limits (Output, Setpoint, Ratio, Bias)
	PV Source, PV Alarming
	Mode Shed on Bad PV
Data Acquisition	PV Source (Auto, Manual, Substituted)
Flow Compensation	PV Clamping
Middle-of-3 Selector	EU Conversion and Extended PV Range
High/Low/Average Selector	PV Value Status and Propagation
Summer	PV Filter (Single Lag)
Totalizer	PV Alarming
Variable Dead Time with Lead/Lag	Bad PV
	PV High/Low
General Linearization	PV HiHi/LoLo
Calculator	PV Significant Change
	PV Rate-of-Change +/-

Figure 6–11   Control unit available algorithms (Re. TDC 3000, courtesy of Honeywell).

The configuration of control schemes by using FBs is carried out by menu selection and filling in the data fields. Normally, there is no need to write computer programs or develop new algorithms; however, it is not difficult to develop new algorithms if required. For instance, it may be useful to combine two or more algorithms into a widely used algorithm to save in documentation, testing, data entry, and so on. In the next three sections, we will show how easy it is to develop some useful algorithms.

6.4.1 PID Algorithm

The PID equation for the controller block diagram of Figure 6–12 is given as follows:

```
        SP                e              C                 PV
    ──►(X)───────►[ PID ]────►[ Process ]────►
        +▲-
         │
         │ PV
         └──────────────────────────┘
```

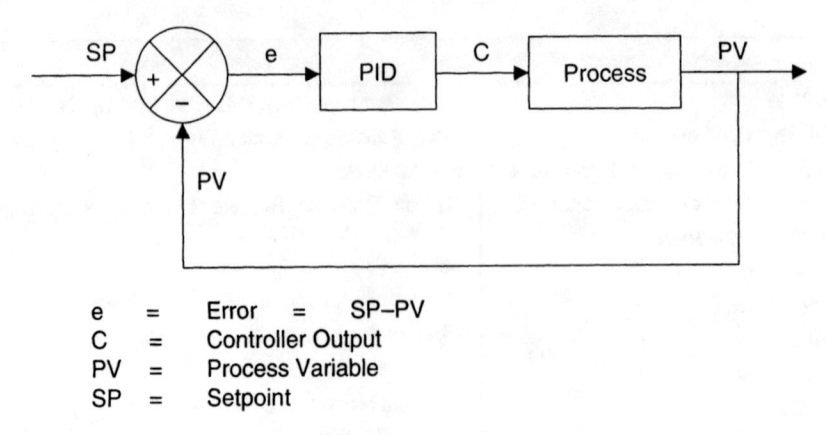

e = Error = SP–PV
C = Controller Output
PV = Process Variable
SP = Setpoint

Figure 6–12 PID control block diagram.

$$C = K\left(e + 1/T_i \int e\, dt + T_d\, de/dt\right)$$

Where K, T_i, and T_d are controller gain, integral time, and derivative time, respectively. Differentiating the PID equation yields:

$$dC/dt = K(de/dt + e/T_i + T_d\, d^2e/dt^2)$$

Using finite difference methods:

$$dC/dt = (C_i - C_{i-1})/\Delta T$$
$$de/dt = (e_i - e_{i-1})/\Delta T$$
$$d^2e/dt^2 = (e_i - 2e_{i-1} + e_{i-2})/(\Delta T)^2$$

Where i is the sample number and ΔT is the integral step length. To solve for C_i:

$$C_i = C_{i-1} + K\Delta T\left((e_i - e_{i-1})/\Delta T + e_i/T_i + \Delta(e_i - 2e_{i-1} + e_{i-2})/(\Delta T)^2\right)$$

ΔT is sampling interval and is normally between 0.1 to 1 second. Rearranging the previous equation:

$$T = \Delta T$$
$$C = C_i - C_{i-1}$$
$$e = e_i - e_{i-1}$$
$$\Delta C = K(\Delta e_i + e_i T/T_i + (T_d/T)(\Delta e_i - \Delta e_{i-1}))$$

This equation can easily be solved if the error e at the present and the last two samples are known. The control unit RAM holds the three error values (and other constants).

6.4.2 Lead/Lag Algorithm

The lead/lag algorithm in Laplace-transform is given by:

$$Y/X = (1 + T_2 s)/(1 + T_1 s)$$

Where Y and X are output and input, T_1 and T_2 are lag and lead times, and s is the Laplace operator. Rearranging and differentiating the previous equation yields:

$$T_1 dY/dt + Y = T_2 dX/dt + X$$

$$T_1 (Y_i - Y_{i-1})/\Delta T + Y_i = T_2 (X_i - X_{i-1})/\Delta T + X_i$$

Assuming $T = \Delta T$ is the sampling interval:

$$Y_i = (T_1/(T + T_1))Y_{i-1} + ((T_1 + T_2)/(T + T_1))X_i - (T_2/(T + T_1))X_{i-1}$$

Assuming $T_1 \gg T$, and $T_2 \gg T$:

$$Y_i = Y_{i-1} + (T_2/T_1)(X_i - X_{i-1})$$

6.4.3 Square Root Algorithm

Interactive methods can be used to solve an equation $f(x) = 0$. If x_i is an approximate value to satisfy $f(x) = 0$, a better value for x can be calculated from the Newton-Raphson equation:

$$x_{i+1} = x_i - f(x_i)/f'(x)$$

Where:

$$f'(x) = df(x)/dt$$

Using the above method to find $x = \sqrt{a}$:

$$x^2 = a, f(x) = a - x^2 = 0$$

$$x_{i+1} = x_i - (a - x_i^* x_i)/(-2x_i) = 0.5(x_i + a/x_i)$$

In the algorithm, an initial value of $x_0 = a$ will be used. For example, for a $= 4 = x_0$:

Interaction 1: $x_1 = 0.5(4 + 4/4) = 2.5$
Interaction 2: $x_2 = 0.5(2.5 + 4/2.5) = 2.05$
Interaction 3: $x_3 = 0.5(2.05 + 4/2.05) = 2.0006$

After three interactions, we have found the square root of 4 as 2.0006. The error of calculations is 0.03 percent.

The main reason behind explaining the development of these three algorithms is to show how easy it is to use computers to solve various control engineering problems. These algorithms (and many more) are readily available in the ROM of the PCS control unit for use, but more complicated or unusual algorithms can be developed by the application engineer. Normally, a simple-to-use process-oriented programming language is available in the CU for such applications.

6.5 INTELLIGENT SYSTEMS

6.5.1 Introduction

The traditional process measurement, control, and monitoring techniques have limitations in the solution of many important problems in control systems engineering. First, the control and instrumentation of process plants has the potential to provide a large amount of data (by field sensors), which can be used for such tasks/functions as process control, process modeling, optimization, diagnostics, and fault prevention. Yet, the inputs processed by control systems are limited by the following constraints:

- Complexity of control algorithms
- Inefficiency of control units to process a vast number of process inputs/outputs
- The long time it takes to process a large amount of process parameters because of hardware limitations
- Ever-increasing demand on control and instrumentation systems because of increasing complexity of chemical processes

Second, the control systems representational and mapping capability is limited by a lack of accurate knowledge about processes and hence inaccuracy of the models used. Process control system models and analysis require accurate data; otherwise, errors by the model can be interpreted as system faults or preventing the detection of actual faults. Systems with the ability to use general plant inputs/outputs (I/O) with accurate models would be highly desirable.

Third, current control systems lack the ability to learn and adapt themselves to process changes. The control units use prespecified algorithms to control processes; however, sometimes chemical processes present unforeseen changes, where their control requires adaptive algorithms with learning capability. Adaptive controllers based on traditional computational techniques to tackle process gain/dead time changes (i.e., nonlinearities) have not found widespread use because of their inability to cope with large process changes. What is needed is the ability to learn by example and search for the solution to a problem.

The traditional process control techniques have reached or are approaching the limit of their capabilities. Consequently, alternative methods are necessary and are constantly being sought by scientists and researchers. An alternative approach with the following features is required:

- The ability to interpret and use the vast quantities of process data
- The ability to respond with high speed to process inputs
- Mapping capability/pattern recognition

- Adaptivity
- The ability to learn and capture knowledge
- Fault tolerant and robust

6.5.2 Artificial Intelligence

In order for an alternative technique to be able to address the shortcomings of conventional control system methods and provide the aforementioned capabilities, it must be able to understand, learn to adapt, and reason logically. It should be able to interpret large quantities of data and recognize patterns in a short time span, which is mandatory in real-time systems. In other words, it should solve problems like a human being. The human behavioral cognition and control can be hypothesized by three levels of behavior: knowledge-based behavior, rule-based behavior, and skill-based behavior.

- *Knowledge-based level*: Where reasoning from stored knowledge or response in an unconscious manner in a feedback-style model is used. This is the equivalent of "human thinking."
- *Rule-based level*: Where in a feedforward fashion, steps are taken to achieve a particular objective. This is equivalent to "human seeing."
- *Skill-based level*: Where control is achieved in a feedforward manner and by using an efficient dynamic model. This is equivalent to "human looking."

A control system based on artificial intelligence (AI) will have to be able to look, see, think, and decide by analyzing a vast quantity of inaccurate and inadequate data. Such systems can learn from historical data. The learning process is analogous to the way in which the human brain learns from experience. The structures of AI-based systems vary; however, the structure of neural networks is believed to be similar to the way the synapses and neurones of the brain are interconnected. Neural networks are discussed in Section 6.5.4. In Section 6.5.3 we examine several techniques currently being applied and studied actively in both industry and academia.

6.5.3 Intelligent System Techniques

A noninclusive list of techniques based on AI is neural networks, fuzzy logic, expert systems, generic algorithms, and qualitative reasoning. Although these techniques have been studied and developed in isolation, it is being recognized that they all mimic the human brain in one way or another. Further, it is gradually being realized that improvements will be made by merging the aforementioned techniques into integrated control systems.

Although fuzzy logic and expert systems have been used commercially in such applications as consumer products (e.g., washing machines, automobiles) and process control (auto-tuning), the neural networks technique is believed to be the most promising solution to process control problems. Section 6.5.4 is devoted to neural networks. In the following paragraphs, a brief description of various AI methodologies is presented.

Neural Networks

Although neural network technology started to attract interest later than fuzzy logic or expert systems, it is probably the first AI-based technique to be discussed by scientists. McCulloch and Pitts introduced a neurone model in 1943. Fuzzy logic and expert systems were introduced more than 20 years later. Neural network algorithms are believed to be by far the most promising among AI-based technologies. Practical algorithms for process control and medical applications have been available since the early 1990s. Neutral networks have the following features:

- Learning capability
- Feedforward model (not feedback)
- Self-organization
- Nonlinear applications
- Function-based (i.e., not hierarchically structured)

Fuzzy Logic

Practical fuzzy logic models were introduced in the 1970s. Fuzzy logic algorithms are widely used in consumer goods (e.g., washing machines, automobiles) as embedded electronic units. It is suitable for poorly defined nonlinear process control applications, such as pH control. It works by using approximate reasoning similar to the human decision-making process when coping with uncertainty and approximation. The main features of fuzzy logic systems are the following:

- Can reason with inaccurate data
- Nonlinear applications
- Simple (soft) programming
- Machine intelligence concept
- Can link with neural network models as an integrated system

Expert Systems

The working of expert systems is based on the experience of a human expert. Considerable interest has been shown in expert systems in such applications as process control and medical and geological prosperity. Foxboro introduced an expert-based PID auto-tuning controller more than a decade ago. Although this debut excited many control and instrument engineers initially, and it was thought that the controller would be used universally, I have not heard much about its success or real applications. This does not imply, however, that expert systems will not be used in the future.

I believe that expert systems will be employed in manufacturing and process control extensively. The most likely configuration will be combined expert systems and neural network models. This will enhance the application domain of both expert systems and neural networks. The main features of expert systems are the following:

- Can reason with uncertainty
- Can acquire knowledge, although not easily, especially deep knowledge
- Can handle nonlinear functions
- Definition/explanation not easy

Generic Algorithms

We are familiar with the shortcomings of quantitative analysis methods, namely limitations on boundaries and finding local optima, rather than global ones. Generic algorithms, which are biologically motivated, endeavor to find global optimum solutions. Although research into generic algorithms has received much interest recently, I expect that a merging of this technique with the other AI methods will take place in order to maximize their application domain and minimize their limitations. The main features of generic algorithms are as follows:

- Search for global optima, not local ones
- Normally not real-time applications
- Handles nonlinear problems
- Multitasking; parallel structure

Qualitative Reasoning

Qualitative reasoning is based on human methods of reasoning. Qualitative reasoning research is mainly targeted into scientific problems; however, merging this method with other AI techniques will offer practical solutions to engineering problems. The main features of qualitative reasoning methods are as follows:

- Reasons with imprecise/inadequate data
- Reasons like a human
- Yields spurious solutions; needs filtering

6.5.4 Neural Networks

Neutral networks are mathematical algorithms that can identify and model relationships between cause-and-effect variables from experience and historical data. The structure of neural network models is similar to the way neurones of the human brain are interconnected. In its simplest form, a neural network is made up of interconnected layers of inputs and outputs. For the brain, the inputs represent information received such as seeing, touching, smelling, and hearing. The outputs (i.e., appropriate response) would be speech, movement, and so on. For a neural network–based process controller, the inputs will be sensors and setpoints, and the outputs will be close, open, start, or stop process equipment. Figure 6–13 shows the structure of a neurone.

Where the neurone output C_j is given by:

122 | INDUSTRIAL PROCESS CONTROL

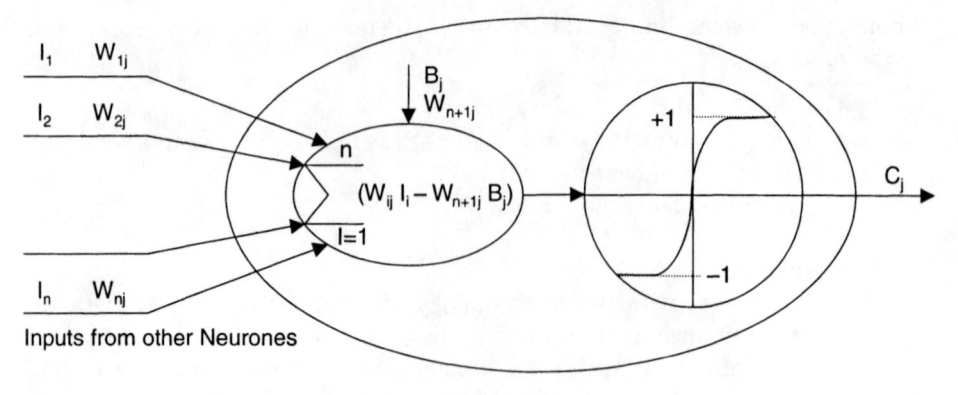

Figure 6-13 Structure of a neurone.

$$C_j = f\left(\sum_{i=1}^{n} W_{ij}I_i - W_{n+1j}B_j\right)$$

The input vector $I = (I_1, I_2, \ldots I_n)$ includes the external and internal (output of other neurones) variables. B_j is an internal bias. $W_j = (W_{1j}, W_{2j}, W_{3j}, \ldots W_{n+1j})$ is the weight factor, which modulates the variables passing through the neurone. The function f is a sigmoid, often given by:

$$f(\eta) = 1/(1 + e^{-x})$$

An alternative function, shown below, has also been used:

$$f(\eta) = \mathrm{Tanh}(x) = (e^x - e^{-x})/(e^x + e^{-x})$$

The function f maps the current state of activation to an output signal. The function f is used by the hidden neurone layers. The I/O layer neurones use linear scaling functions, although in some applications, the output layer neurones may use the sigmoid function.

Neural networks used in process control are multilayer feed forward–type models. Multilayer feed forward neural networks are used in process control because they can approximate nonlinear processes with an acceptable level of accuracy. They can also be trained by learning from the control systems historical databases. Figure 6-14 indicates the structure of a multilayer feedforward neural network.

The calculations within a neurone layer may take place asynchronously (parallel process); however, interaction between adjacent layers is synchronous. Training of neural networks involves finding optimum weights (W matrix). The training algorithm can be based on back propagation methods, where from known inputs and outputs the weights and the number of hidden layers are calculated. The output C_{ij}, for the *i*th neurone of *j*th layer is given by:

$$C_{ij} = f\left(\sum_{m=1}^{n} W_{mij}C_{mj-1} - W_{n+1j}B_j\right)$$

(The number of hidden layers could be one or more)

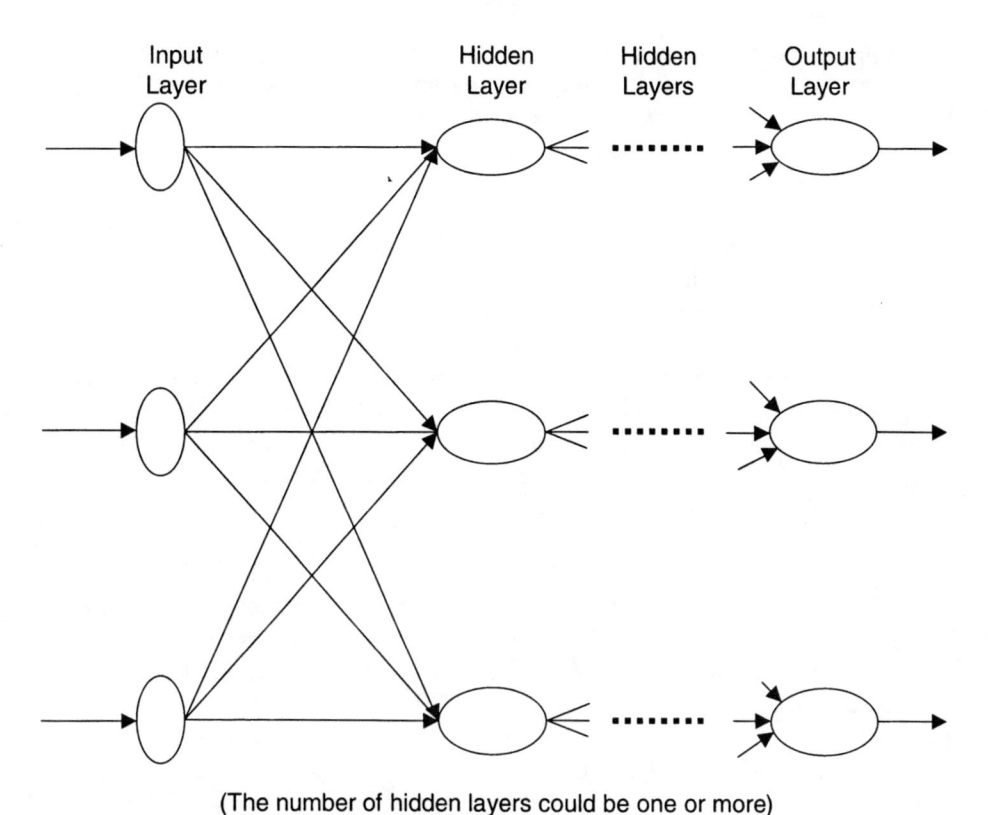

Figure 6–14   A multilayer feedforward neural network.

Where n = the number of neurones in the *j-1*th layer, W_{mij} = weight for the *m*th input of the *i*th neurone of the *j*th layer, C_{mj-1} = the output of the *m*th neurone in the *j-1*th layer.

The neural network model described above requires historical data, not a model with known dynamics or an analytical objective function. It does not assume any predefined control law; it learns the control law autonomously. It relies on reinforcement learning to not only model the process but also to develop a strategy to control the process. Neural networks can use large quantities of field sensor data. Adding more inputs to the network affects neither the computational complexity nor the model configuration.

Neural network activities include data collection, data processing, parameter selection, training, and validation. These functions will be briefly explained in the following paragraphs.

Data Collection

The existing control and instrumentation systems are capable of providing the necessary data for neural networks. They can scan and store a large quantity of process parameters (inputs, outputs, setpoints, and diagnostics). The plant PCS can be equipped with the required mass memory (disk, tape) to hold historical data needed by the neural network. The application engineer, with the help of the neural network itself, will decide which items of data are useful for the neural network model.

Data Processing

The raw process data have to be conditioned or preprocessed before a neural network can utilize them for training. Raw process data normally contain out-of-range values, missing values, corrupted values (e.g., due to noise), or unrepresentative values (e.g., invalid data). Statistical methods are used to replace missing data, out-of-range data, and erroneous values. Such techniques as regression, partial least square, and cross-correlation can be employed to recover missing data. If the available historical data were large enough, then a simple way to treat missing or out-of-range data would be to delete them from the system.

Parameter Selection

The present-day C&I systems collect so many process plant parameters that it is debatable whether a large portion of them will ever be utilized. In a recent offshore complex, where I was responsible for the design of the control and safety systems, the PCS had more than 20,000 process inputs/outputs, and the mechanical/electrical packages control systems had approximately 50,000. In a large process plant, the C&I systems may cater for more than 100,000 parameters, if we include system diagnostics and equipment status variables. Many parameters will be stored in the control system host for varying lengths of time—a few hours, days, weeks, and years.

It is obvious that if we employ one or several neural networks in a large plant, we will use only a very small portion of the C&I parameters. A typical neural network may use 10 to 100 inputs/outputs and one to three hidden layers. Adding more inputs or hidden layers may increase the complexity and processing time of the neural network substantially, while not improving its performance. Only the variables that are important for the neural network should be selected.

Various statistical methods, such as regression analysis (principal component analysis), linear regression, and sensitivity analysis, can be used to eliminate the parameters that do not contribute to the outcome of the neural network model. Two methods of parameter selection are available: formal selection and backward elimination. In the forward selection method, we start with a few parameters, which we know are important for the model. Then we add more variables and measure their effect on the neural network outputs. If they improve the model performance, we include them in the calculations; otherwise they are discarded. In the backward elimination method, all variables are included in the neural network model initially. Then the parameters that are not relevant are deleted.

An important subject when selecting neural networks inputs is the process dynamics (i.e., time constants and dead times). As neural networks measure the process dynamics, all of the process parameters, which are necessary to measure dynamics, should be included in the model. If time constants and dead times are known, they can be fed to the model as inputs.

Training

Setting the weights and biases requires the model to be trained, which is the neural form of programming. The model is presented repeatedly with a selection of examples of situations it may encounter, and letting it know the correct outputs. Training/test data are used by the training algorithm to find optimum weights and biases. A good training algorithm will produce results within an acceptable time scale and avoid overtraining. Overtraining means using excessive iterations and/or hidden layers to train the model. The adverse effect of overtraining is the inability of the neural network to generalize on new data. A set of test data is normally used to justify the training and avoid overtraining.

Two types of training are available: supervised and unsupervised. Supervised training is suitable for PCS because it can handle nonlinear processes more efficiently than the unsupervised type.

Validation

Validation means determining when the neural network model produces reliable outputs and when it does not. Neural networks will not perform viably in areas where they have not been adequately trained. A validity measure should be included in the model to indicate the reliability of the neural network prediction. This will indicate when the calculations are not reliable. Statistical methods may be used to find validity measures.

Use of Expert Systems

In order to optimize the aforementioned functions (data collection, processing, selection, training, and validation) and continually improve the performance of the neural network, the use of an expert system is highly recommended. An expert system can manage every step in building a neural network model. It can, for example, advise what type of statistical analysis is suitable for a particular calculation, when retraining or testing is required. Figure 6–15 shows a combined expert system and neural network model.

Neural Network–User Interface

It is imperative that the neural network system should provide a user-friendly environment. Application engineers with little or no software experience should be able to install, configure, and apply neural networks. The user may wish to monitor and control the neural network training and testing or leave them to the system. It should be possible for the application engineer to transfer his or her process knowledge and experience to the neural network. The neural network should be able to use the process data even if the information is incomplete, inaccurate, or accompanies tolerance bands. Such tools as menu-driven displays, windows, spreadsheets, and graphics are normally provided.

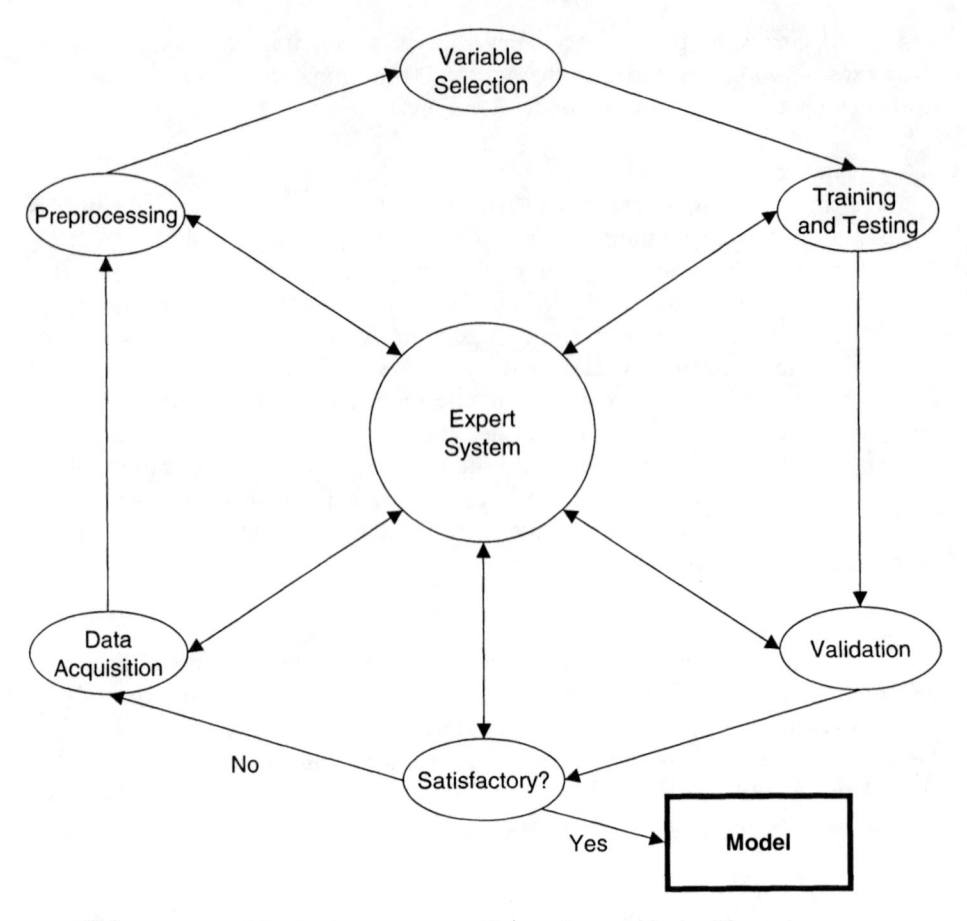

Figure 6–15 A combined expert system and neural network model.

Neural Network–PCS Interface

The present-day control systems are open systems and normally use UNIX or Windows NT operating system and Ethernet LANs. Therefore, if the neural network is an open system, it will be easy to interface or even implement the model in PCS (e.g., in PCS host). In such a system, if the neural network controls some parts of the process, it should be used in a supervisory control mode, rather than direct control; however, if the neural network algorithm resides in the PCS control unit, it may be used for direct control, but this is not recommended. PID control schemes need fast and tight feedback control. It cannot be guaranteed that current neural networks can satisfy this type of control.

Conclusion

As the preceding paragraphs indicate, the application of neural networks is much simpler than the conventional process modeling methods. The latter requires a deep knowledge of process dynamics and advanced mathematics/ simulation. Neural network methods rely on simple algebra and operator experience. Unlike optimization/simulation algorithms, neural networks are computationally efficient. Neural networks can be trained, retrained, tested, and improved similar to the way people gain experience and learn to do things. Neural networks can work satisfactorily, even though the model inputs (training or working) may be incomplete or inaccurate. This is not the case for orthodox control algorithms.

6.5.5 Application Example 1

Consider the control of an irregularly shaped tank, as shown in Figure 6–16. This is obviously a highly nonlinear process, which will be difficult to control with conventional PID controllers. Nonlinearity caused by abrupt transitions between narrow and wide sections can be easily handled by a neural network–based controller. Such a neural network controller can be easily trained by providing it with a few examples. The rates of change of level at transition points and three different area points versus incremental control valve opening

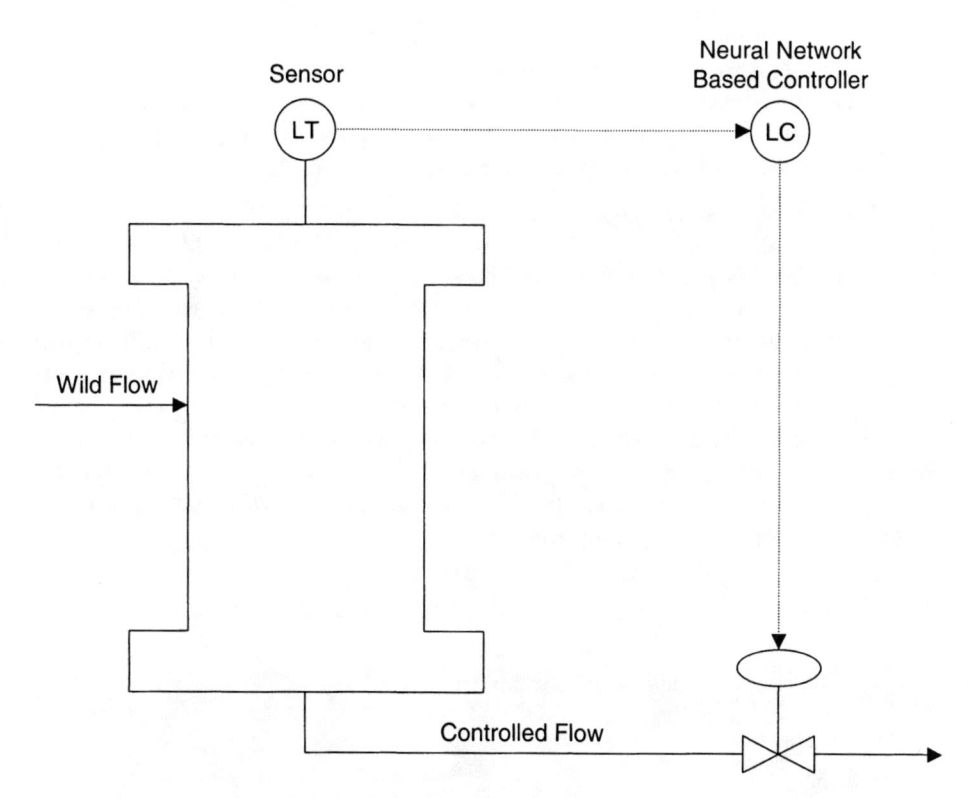

Figure 6–16 Control of an irregularly shaped tank.

are adequate to train the controller. Such a controller has been successfully developed and tested at AI Ware Inc.[16]

6.5.6 Application Example 2

Fisher-Rosemount[12] has developed an intelligent sensor based on advanced neural network technology. Many well-known process parameters cannot be measured directly or accurately by traditional instrumentation and measuring devices. Some variables can be measured accurately, although the measuring equipment/methods may have the following deficiencies:

- Long response time; normally not acceptable for real-time process control
- Large equipment size
- Prohibitive equipment cost
- High cost of maintenance and operation

The Fisher-Rosemount sensor has been applied to some difficult measurements in pharmaceutical and pulp and paper industries, and the results have proved its superiority over conventional measurement techniques. Fisher-Rosemount has successfully completed field tests for the following applications:

- Viscosity measurement in a chemical plant
- Temperature anomalies in a batch reactor
- Pulp brightness in a recycled paper mill
- Particle size and flowability of powder in a soap factory
- Melt index of high-density polyethylene

The sensor uses historical process data for training and can be configured in an existing PROVOX or RS3 control unit (e.g., IFC, UOC, MPC) or other control systems. The inputs to the intelligent sensor will be from the CU database. If the existing process variables are not adequate for the sensor to provide accurate measurement, additional field transmitters may be installed. This will probably be unavoidable in existing plants, but normally the installation of a few transmitters is not a difficult undertaking.

For a new plant, it is advisable to seek help from Fisher-Rosemount to ensure that all necessary process parameters are available for the satisfactory operation of the intelligent sensor(s). The following list shows some applications for the Fisher-Rosemount intelligent sensor:

- Brightness (pulp and paper)
- Bubble size (chemical)
- Color
- Column flooding/foaming (refinery)
- Composition
- Density
- Efficiency
- Emission

- Fermentation rate
- Flavor
- Flowability
- Kappa
- Moisture
- Molecular weight
- Octane number
- Odor
- pH
- Paper thickness
- Particle size
- Specific gravity
- Time left to complete a reaction
- Viscosity

6.6 SIMULATION

6.6.1 Introduction

The purpose of simulation is to study the behavior of a system by observing the response of a model of the system to changes during a period of time. Although physical models can be used to simulate a system, only mathematical models will be considered in this book. Simulation packages that can be used to simulate any type of process are available in the market. The use of such software packages is easy and does not require any software expertise.

The following steps are normally taken in the development of a system model and simulation:

1. Define the problem.
2. Plan the study.
3. Build a mathematical model of the system.
4. Use a software/database package to program the mathematical model.
5. Validate the model.
6. Design experiments.
7. Run the software program and analyze the response.

The accuracy of the simulation model is important and will be dictated by the application. The application of the simulation program may be for:

- Training
- Control
- Design

A high degree of accuracy is required for simulation models used for the design of process equipment and plants. The mathematical model should be accurate, and the data used for simulation should be highly reliable and representative of the real system. On the other hand, a training model and the data used for such simulation do not need to be as accurate as those used for design and operation.

130 | INDUSTRIAL PROCESS CONTROL

Although special simulation packages are normally necessary for the design and operation of process plants, application engineers may employ the PCS control unit algorithms to develop small simulation models for training or process control. In the following two sections, we will see how simple simulation models that are suitable for training and improving process control can be built.

6.6.2 Simulation and Process Control

Figure 6–17 shows the piping and instrument diagram (P&ID) for a heater system. It is possible to improve the performance of the heater operation by using simulation. Heaters, especially large ones, have two components that can be highly nonlinear and contribute to poor control. The two components are (1) first-order lag (T_1), which may change with time or process conditions, and (2) pure dead time (T_2), which may vary with process conditions.

By simulating first-order lag (T_1), or dead time (T_2), or both, it is possible to improve the control performance. Figures 6–18 and 6–19 show the block diagrams for such simulation control schemes.

The dynamic feedforward control (Figure 6–18) has four components:

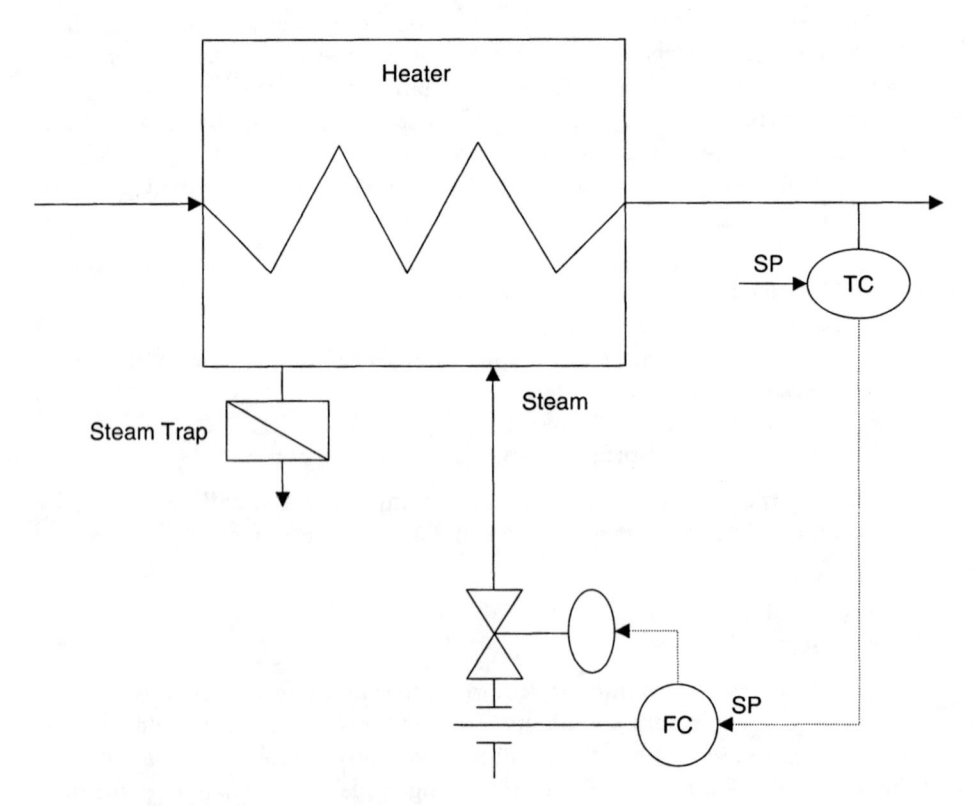

Figure 6–17   Heater system P&ID.

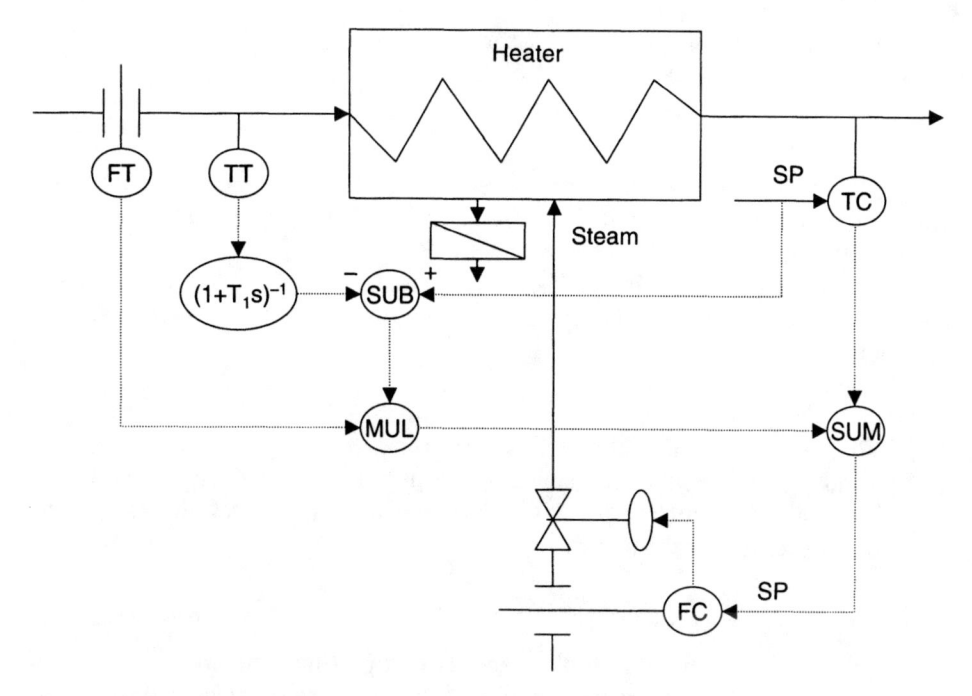

Application Engineering of Control Systems | 131

Figure 6–18   Heater control with first-order lag T_1 simulation.

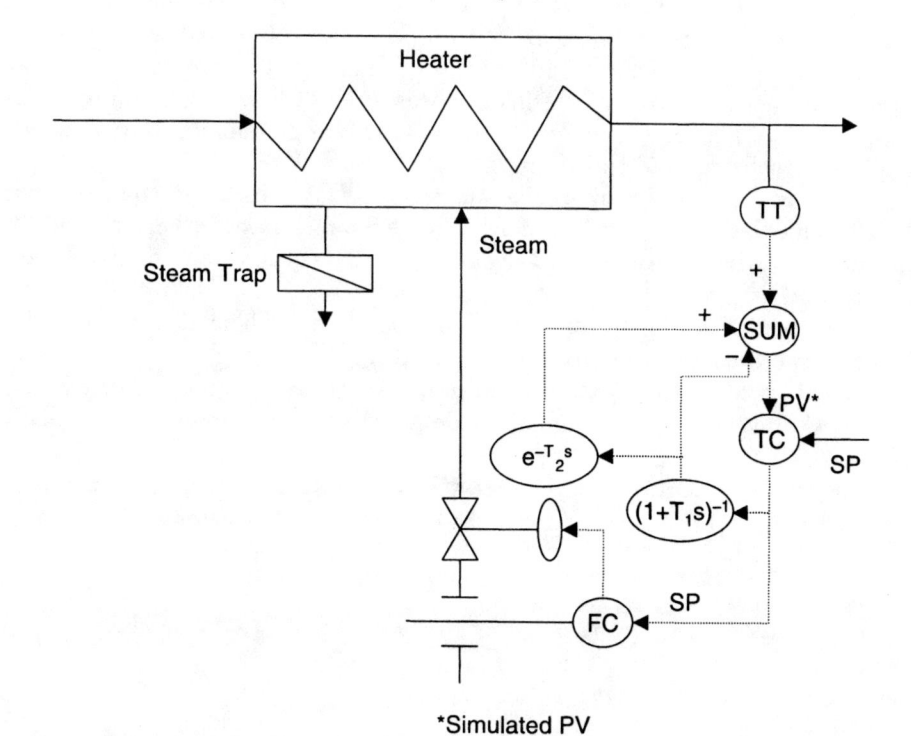

Figure 6–19   Heater control with dead time T_2 simulation.

- A dynamic lead/lag compensator
- A demand calculator
- A feedback PID controller/trimmer
- A fuel (supply) PID controller

The lead/lag algorithm compensates for the process exponential lag. The output of the demand calculator is trimmed by the TC output and then is used as the setpoint by the fuel controller (FC).

The Smith Control Scheme (Figure 6–19) will compensate for the time delay (dead time) of the heater. The process is represented by the following transfer function:

$$e^{-T_2 s}/(1 + T_1 s)$$

Where T_1 = exponential lag, T_2 = dead time.

Dead time is a nonlinear element and can destabilize the control loop. The use of the Smith Control Scheme will reduce the possibility of destabilization and improve the control performance.

6.6.3 Simulation and Training

Training of engineers and plant operators by using simulation is common practice. Before the introduction of computer-based control systems, physical and analog simulation methods were employed. Since the introduction of computers, especially microprocessor-based DCS, to process control, digital simulation has become the most popular method of training in all types of process plants. The simulation method described in this book uses the PCS standard control algorithms. It does not require in-depth knowledge of programming, control theory, Laplace–transform, or Z-transform.

The first step in developing a simulation model is to break down the process (P&ID) into blocks. A block could be, for example, a valve, heater, tank, T-junction, or process dead time. Typically, a block will have one or two inputs and an output.

The second step is to translate each block to a mathematical model. This is the most critical step, and for accurate modeling, a thorough understanding of the principles of thermodynamics and unit operations is required. For training purposes, however, an approximate model, as described as follows, is normally adequate.

It can be assumed that a process block consists of two components, a first-order lag (T_1) and a pure dead time (T_2). This can be represented in Laplace–transform as:

$$X \longrightarrow \boxed{T_1, T_2} \longrightarrow Y$$

$$Y = \frac{kXe_2^{-Ts}}{1 + T_1 s}$$

A valve can be replaced by a first-order lag, two first-order lags, or a first-order lag and a dead time (for large valves). A tank can be represented by an integrator (1/Ts). A heater can be simulated by one or two first-order lags, or by a first-order lag and a dead time. In large heaters with long pipelines, the dead time may change with flow. This can be incorporated into the model. Three examples—a tank, a heater, and a binary distillation tower—are modeled here. Other processes can as easily be modeled. Figure 6–20 shows the P&ID for the tank process.

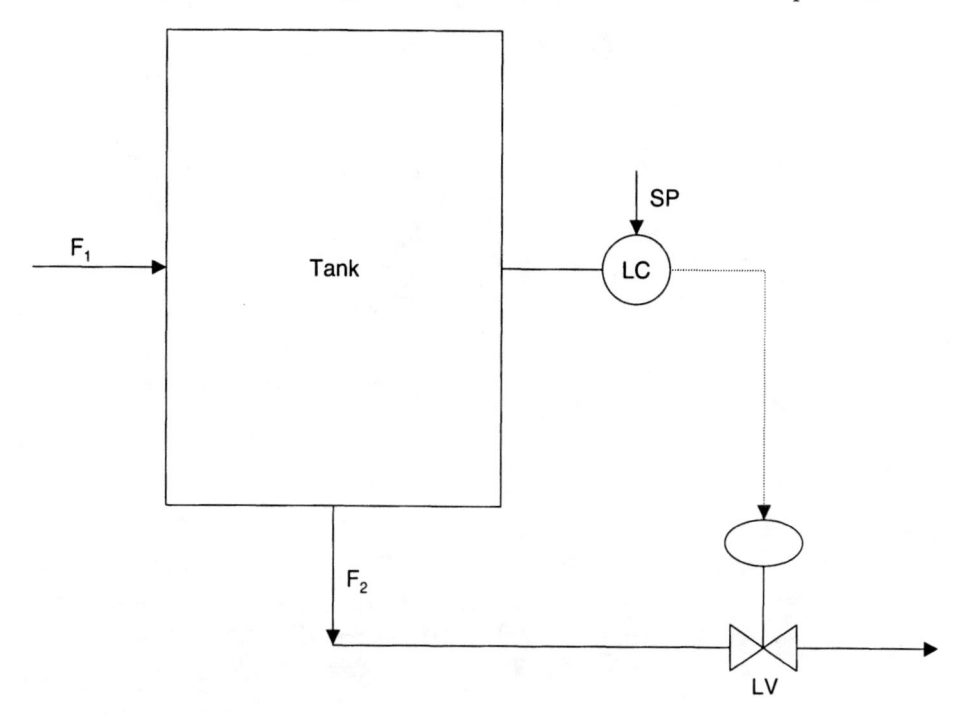

Figure 6–20 Tank process.

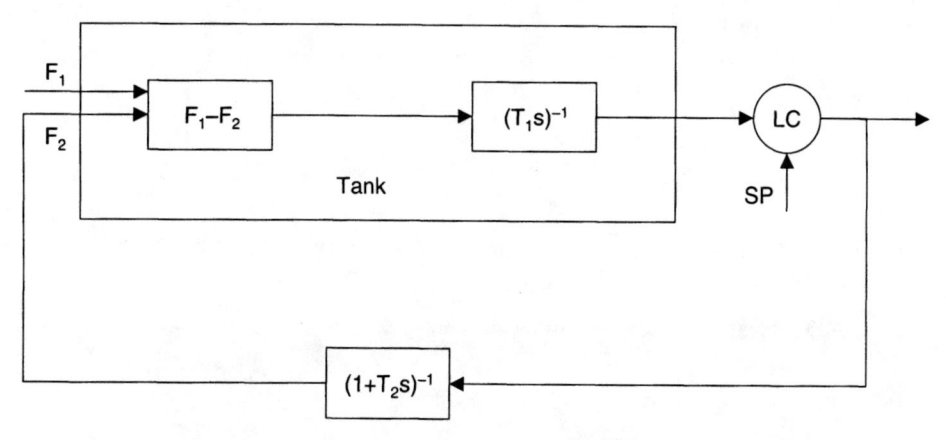

Figure 6–21 Simulation block diagram for the tank process.

134 | INDUSTRIAL PROCESS CONTROL

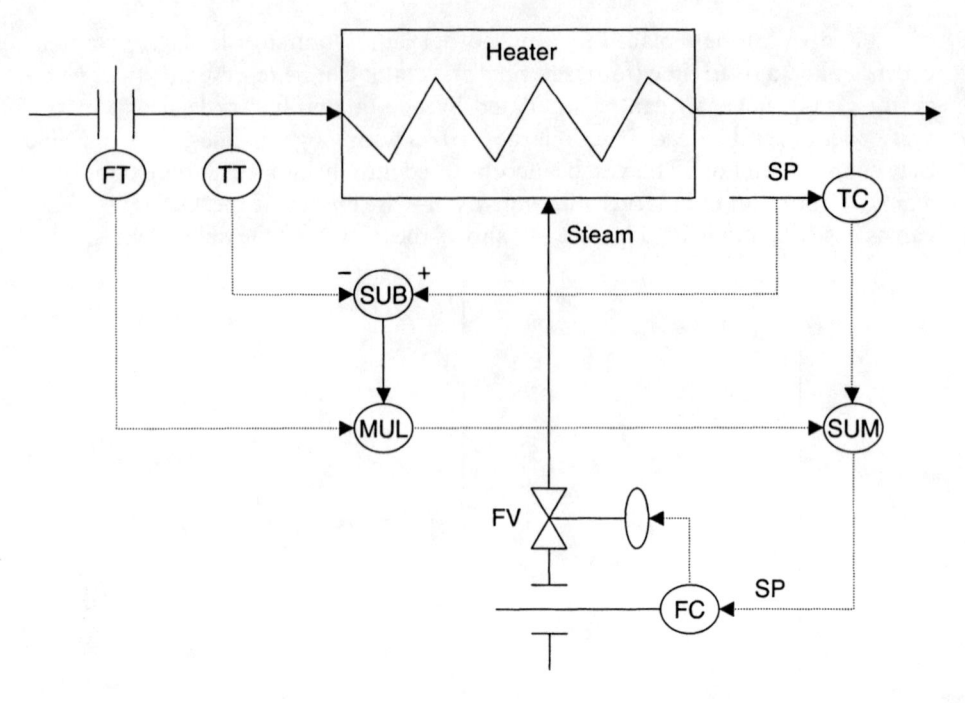

Figure 6–22 Heater P&ID.

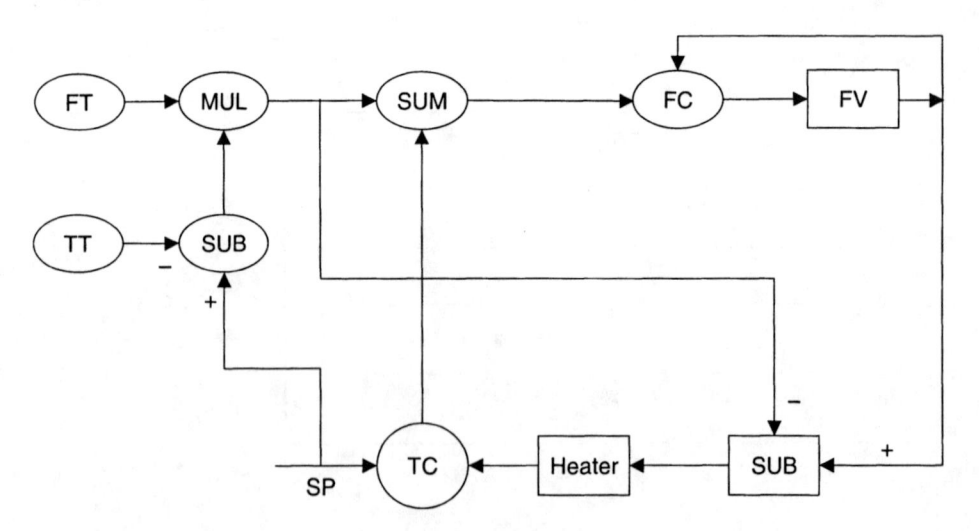

Figure 6–23 Heater simulation block diagram.

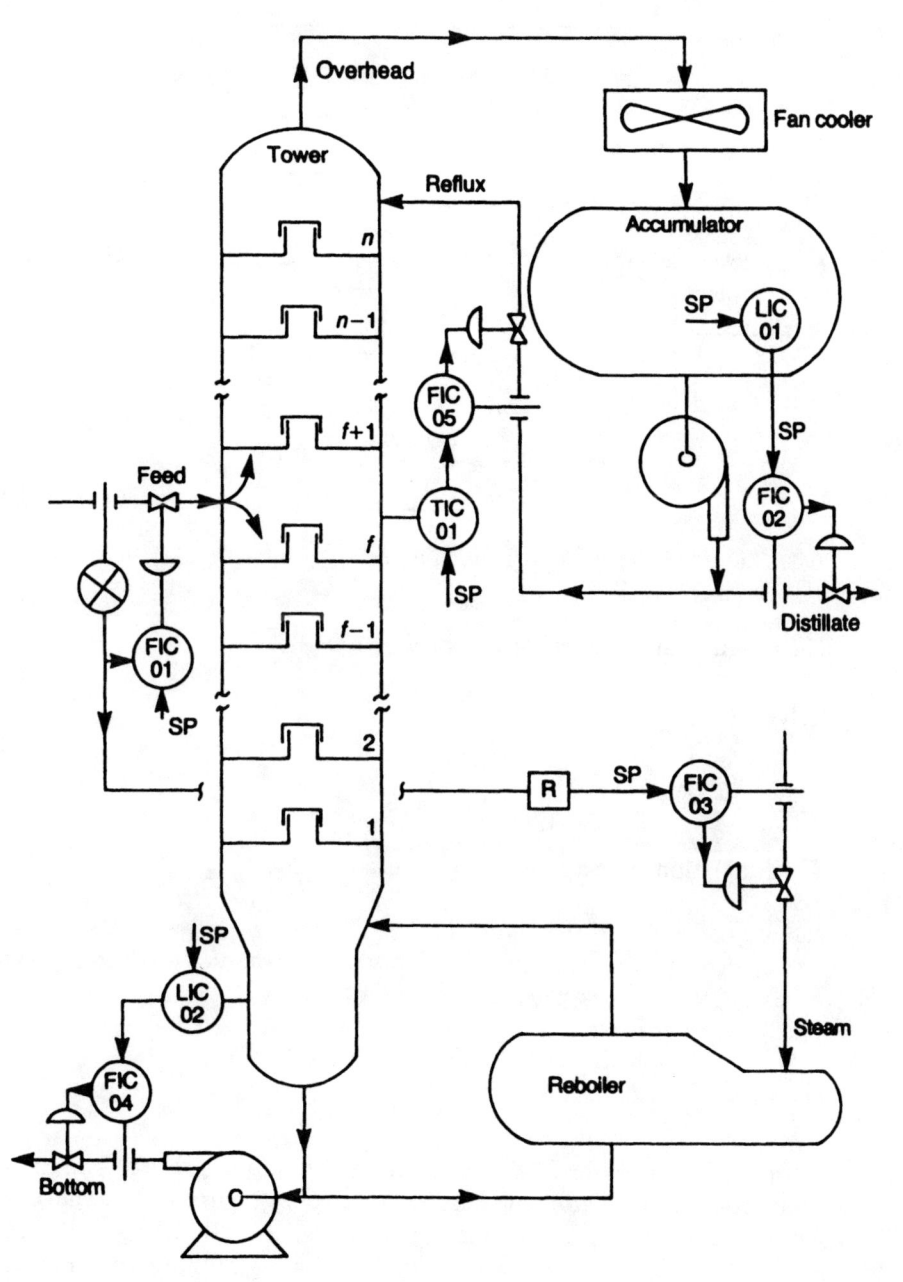

Figure 6-24 Distillation tower P&ID.

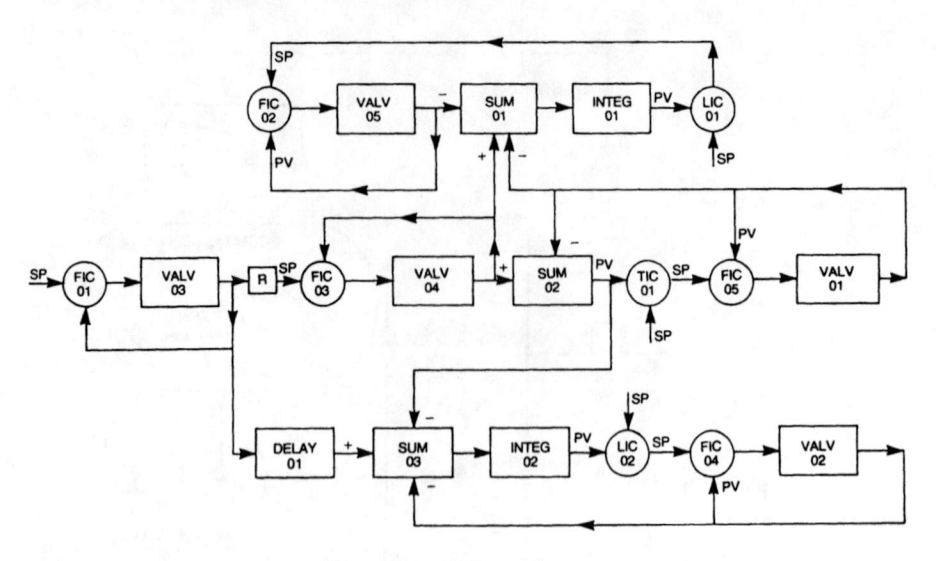

Figure 6–25 Distillation process simulation model.

The required algorithms for this process are:

Valve: $\dfrac{1}{1+T_2 s}$

Level: $\dfrac{F_1 - F_2}{T_1 s}$

The simulation block diagram for the tank process is indicated in Figure 6–21.

Figures 6–22 and 6–23 show the P&ID and simulation block diagram for a heater process, respectively. The algorithm for the valve is shown in the previous example (the tank process).

The heater algorithm is:

Heater: $\dfrac{e^{-T_1 s}}{1+T_2 s}$

Figure 6–24 shows the P&ID for a binary distillation process. The simulation model for the distillation process is depicted in Figure 6–25. The model includes several valves, tanks, summers/subtractors, and PID controllers. I have built this model and carried out various experiments (e.g., tuning the controllers, injecting disturbances). It is a useful simulation model, especially as a training tool. (If a more accurate simulation model is required, see references 7 and 17.) Software packages are also readily available to help with building simulation models for such processes.

7

Some Typical Control and Safety Systems

7.1 INTRODUCTION

In this section, six control and safety systems supplied by large multinational companies are reviewed. The omission of others does not mean that they are less capable or less popular than those included here. The main reasons for studying these six systems are (1) my familiarity and experience with them; (2) because they have been widely applied in the North Sea offshore installations; and (3) because some of the systems offer unique features (e.g., true system integration, open system technology) that place them well ahead of their competitors. Covering fewer systems will also allow us to study them in greater detail. Each system will be studied with reference to the capabilities and features for the following components:

- Internal communication (LAN, I/O bus, etc.)
- Operator interface
- Control unit(s)
- Host computer
- External interfaces (serial links, fieldbus)
- Application packages

The information in this section is based on documents provided by the vendors and discussions held with their application engineers. Because the vendors enhance the capability of their systems frequently, up-to-date information should be obtained from the system suppliers. It is possible that some of the descriptions presented here may be superseded by new information before this book reaches readers.

7.2 ABB MASTER

7.2.1 An Overview of ABB Master

Asea Brown Boveri (ABB) acquired Taylor Instruments and Kent Process Control Limited in the late 1980s. The two companies were supplying their popular distributed control systems (MOD 300 and P 4000) at the time of takeover. Since the introduction of ABB Master in the early 1990s, MOD 300 and P 4000 have been withdrawn from the market. Although if some existing satisfied users of MOD 300 and P 4000 insist on procuring these systems—most probably for expansion rather than totally new systems—ABB will oblige its customers. MOD 30 (MOD 300's predecessor) and P 4000 were studied in my first book.[1]

ABB Master[18] is a true integrated control and safety system and conforms to open system interconnection (OSI) standards. The system consists of several powerful microprocessor-based control modules, which can be programmed (application software) to fulfill the requirements of the following applications:

- Machine control
- Batch control
- Logic/sequence/regulatory control
- Emergency shutdown (ESD)
- Fire and gas detection (F&G)
- Subsea master control station (MCS)
- Heating, ventilation, and air conditioning control (HVAC)
- Drilling control and monitoring
- Vessel control and monitoring (VCS)

Although ABB Master is a relatively new system (compared with MOD 30, TDC 2000, or PROVOX), it has been recognized as a true integrated system and has been selected for several prestigious North Sea installations. With ABB's vast experience and involvement in power plants and electrical switchboards, and the fact that they own some design, engineering, and consulting companies, ABB Master will pose a real challenge to other big PCS vendors.

7.2.2 ABB Master System Internal Communications

The ABB Master system communications comprise a set of database/links that are designed in compliance with ISO standards. The system configuration not only yields an open system architecture but also makes the task of expansion extremely easy. Changes in system topology or adding extra facilities to the system, for example, after the completion of the design, will pose no problem. Figure 7–1 shows the ABB Master system configuration and various communication subsystems.

ABB Master communication subsystems include:

- Master Bus 300
- Master Bus 300E
- Master Bus 200
- Master Bus 90
- Master Fieldbus

Master Bus 300 is a high-performance LAN, which conforms to the IEEE 802.3 (Ethernet) standard. The communication protocol is carrier sense multiple

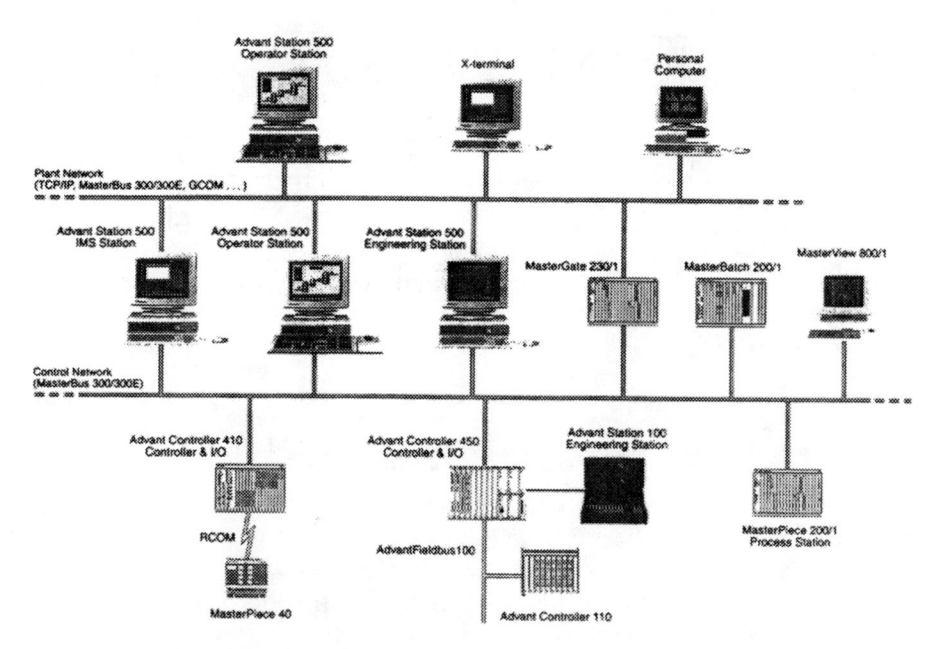

Figure 7-1 ABB Master system configuration.

access/collision detection (CSMA/CD) medium access control, in accordance with ISO Class 4 standard. It is a democratic protocol (i.e., all stations have equal access to the bus). Some PCS vendors who use Token Bus LAN (IEEE 802.4) may claim that Ethernet is not reliable for process control because there is no guarantee that all stations will be granted access when required; however, this claim is not valid for the following reasons:

- Ethernet LAN is extremely fast; 10 MBaud at present and 100 MBaud or higher rates in the near future will be available.
- Data exchange in process control LANs does not normally need high speed.
- Communication protocols normally use some control routines to ensure a reliable flow of data.

Master Bus 300 is a multidrop LAN with up to five segments, each with up to 500 meters. Repeaters can be used between segments to increase the overall length of the LAN. Coaxial and or fiber-optic cables may be used, although the latter needs opto-couplers. Where several networks have to be interconnected to provide a plantwide LAN (or wide area network [WAN]), the ABB Master Bus 300E will have to be used.

The Master Bus 300 transmission speed is 1 MBits per second. This yields a data transfer rate of approximately 1,500 messages (packets of information) per second. The length of each message can be up to 256 Bytes.

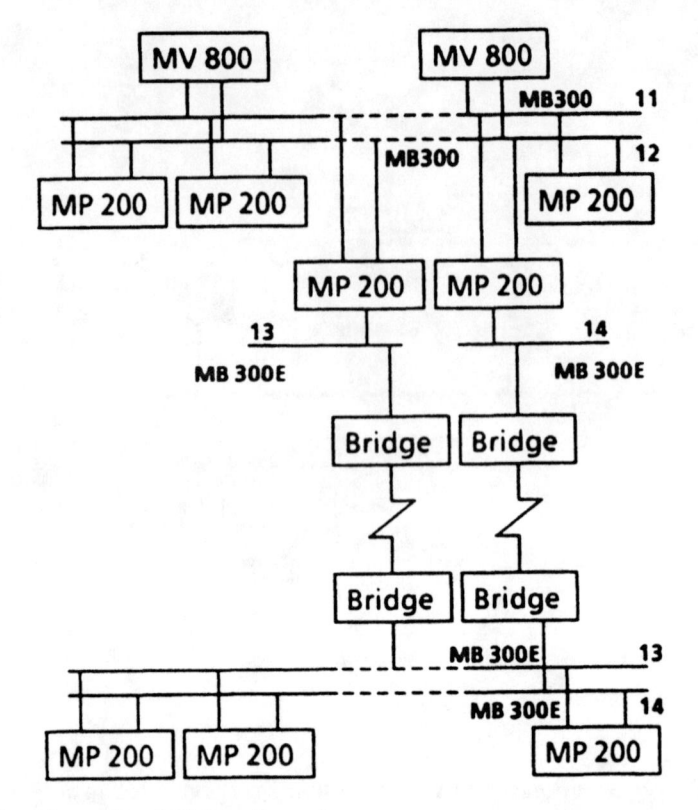

Figure 7-2 Master Bus 300E configuration.

Master Bus 300E (Extended) is used when the PCS LAN has to cover a large geographical area (e.g., as a WAN). To facilitate the extended coverage, it employs bridges for interfacing via radio link and satellite. Figure 7-2 shows a typical application for the Master Bus 300E.

The Master Bus 300E characteristics (medium, protocol, etc.) are similar to those of the Master Bus 300. The performance of the bus depends on the configuration (bandwidth, delays) of the bridges used.

Master Bus 200 is suitable for small systems where the number of stations and the amount of data exchange do not warrant a full ABB Master Network (300 or 300E). The medium can be twisted pair or fiber-optic. It can be used as a local bus in a point-to-point or multidrop configuration. A Master Bus 200 can handle up to 10 stations. A Master Bus 200 can cover distances up to 7 km, if repeaters are employed.

The Master Bus 200 speed is 153.6 kBaud. A data transfer rate of up to 5 kBytes per second can be achieved (depending on the bus configuration). A typical system configuration using Master Bus 200 is a Master View 800 Operator Station with one to five Master Piece Controllers in a multidrop fashion. For remote communication and interface to foreign systems, refer to Section 7.2.6.

Master Bus 90 is a high-performance bus that facilitates data transfer between control units in a cyclic mode or by event (i.e., event driven). Cycle time of 1 msecond or slower can be selected (1, 2, 4, . . . 4,096 msecond). The communication speed is 1.5 MBaud.

Master Fieldbus is used to link remote input/output (I/O) units to a Master Piece 200 control unit. Up to four Master Fieldbuses can be connected to a control unit. Twisted pair, coaxial, and fiber-optic cables with a maximum length of 3,000 meters are available. Master Fieldbus has a transmission rate of 2MBaud.

7.2.3 ABB Master Operator Interface

ABB Master View equipment includes inexpensive monochrome VDU terminal, Advant Station 100 (Lap Top), Advant Station 500, and Master View 800. These operator stations provide full control and monitoring facilities for efficient operation of process plants.

The Master View 800 has standard/popular displays, such as overview, group, detail, point, trends, and diagnostics. The display hierarchy is flexible and configurable, which implies that the user can define a hierarchy system based on process requirements. Display selection is via menus and configurable keys. By the use of configurable keys and hierarchy configuration, useful links between related displays could be created, which will help operators during normal and emergency cases. For example, moving from overview to a process display and/or a group display and vice versa can be accommodated easily.

Master View alarm/event handling includes standard means (e.g., color, priority, flashing, acknowledge, and reporting). Additionally, it enables group alarm management, where several alarms can be grouped together for joint alarm processing. For instance, alarm disable, enable, acknowledge, outputting, and blocking can be applied to a particular group of alarms. This feature can be of special assistance to operators to block an area of the process from being processed for alarm display and action.

The ABB Master person–machine interface facilities include Master View 800, Advant Station 500, Advant Station 120, PC, and X-Terminals. Master View 800 and Advant Station 500 communicate with process control stations (Master Piece 200, Advant Controller 400, etc.) via the control network (Master Bus 300 and 300E). Because the ABB Master View 800 and Advant Station 500 together with the control network yield an open system environment, it is easy for foreign computers to interface with ABB Master. Of course, the former computers (or systems) must conform to OSI standards. The ABB Master Stations use UNIX operating system, X-Windows, SQL-Compliant Relational Database, User API, which in conjunction with ABB Master Network IEEE 802.3 standard, makes application software development and interfacing to other systems highly efficient and trouble-free. Foreign systems running on UNIX, MS-DOS, MS-Windows, OS/2, and VHS can easily interoperate with ABB Master.

ABB customers who are currently employing MOD 300 or P 4000 can upgrade their systems by installing the ABB Master Advant Stations 500 and the plant network. This will improve operator interface and information management (reports, event handling, etc.). Because of the variety of ABB Master Operator interface facilities, plants of any size—from very small, with a few hundred inputs/outputs, to very large, with tens of thousands of inputs/outputs—can be accommodated easily.

7.2.4 ABB Master Control Units

ABB Master Control Units include Advant Controllers 500, 400, 100, and Master Batch 200. These control units interface operator stations via the ABB Master Control Network. They can also interchange data among themselves over the Advant Fieldbus 100. This feature is useful when control units in different parts of a plant need to exchange data (e.g., for cascade control, ratio control, interlocks, scheduling) to cater for a large control scheme. This data exchange over the Fieldbus 100 will avoid hardwiring between control units.

ABB Master Control Units provide standard I/O cards to handle all process requirements (e.g., analog inputs/outputs, digital inputs/outputs, pulse inputs up to 2.5 MHz, variable speed drives, and load cells for weighing). To reduce the field cabling effort, remote I/O card files can be distributed in the field and connected to the control units via ABB Master Fieldbus.

ABB Master Control Units use an easy programming language for the development of application database and control functions. It uses function blocks and graphical representation, which are especially oriented to process control applications. A wide range of ready-to-use algorithms (functions) (e.g., PID, OR gate, AND gate) are available. The developed application programs can be documented graphically by the use of ABB Master Aid (a laptop PC-based package). Master Aid additionally checks the application program for syntax integrity. It can also be used for testing and fault finding of programs and control schemes.

Master Batch 200 is specifically designed for batch control. It provides all standard batch control functions, such as recipe handling, stock control, production planning, batch routing, and report generation.

Master Piece 200 is provided with necessary hardware (thruster interface) and software for vessel control and management. The system offers the following features:

- Dynamic positioning system
- Manual thruster control
- Auto sailing
- Weather vaning
- Position mooring functions
- Power optimization and management
- Bilge/ballast control
- Machinery control

- Mathematical models for vessel dynamic behavior prediction
- Online consequence analysis for thruster failure and line break
- Offline simulation for operator training

7.2.5 ABB Master Host Computer

ABB Master provides VAX or PC-type supervisory computers. They interface the system via the ABB Master LAN. The hosts support many useful application packages, such as long-term archiving, reporting, optimization, simulation, production planning, and information management. Application engineers can also write their own programs with access to the control system database.

7.2.6 ABB Master External Interfaces

Interface to foreign systems is possible either via the control units (Master Piece 200) or via the system LAN by the use of ABB Master Gate 230. The control unit communication protocol (EXCOM) is an easy-to-use package that supports serial links based on RS 232 and ASCII. Development of nonstandard protocols is also possible by using a special EXCOM software toolbox.

Master Gate 230 has been designed for applications where foreign computers (minicomputers) need to interface ABB Master to collect data for information gathering, optimization, planning, and so on. The interface can be either Ethernet (IEEE 802.3) or a serial link (RS 232, RS 422).

The ABB Master control unit has provisions for interfacing fieldbus and smart transmitters where it is required. The use of fieldbus will significantly reduce the burden of hardwiring, cabling, and marshalling. It will also enable remote diagnostics and calibration of field instruments.

7.2.7 ABB Master Application Packages

ABB Master Piece Language (AMPL) is a function block–type configuration tool with graphical representation. All necessary algorithms for process control (continuous and logic) are available. Function blocks have inputs and outputs and utilize the available algorithms to satisfy most control/logic requirements. Programming can be carried out online or offline for complex applications. Function block inputs/outputs are connected to other blocks or to process I/O as required. The final program can be documented graphically by the use of ABB Master Aid.

Electronic Weighing is incorporated into ABB Master and can interface both standard strain-gauge load cells and ABB's patented Pressductor load cells. ABB has developed special interface cards for both types of cells. Logic and process control is provided via ABB Master Piece Language (AMPL), monitoring, and operator control via operator stations.

Variable Speed Drive Control is provided via specially designed master piece boards or via master fieldbus. In the latter method, the ABB AC and DC drive controllers (SAMI and TYRAK) are connected directly to the master fieldbus and the fieldbus to master piece. In the former case, a pulse counter in the interface board accurately measures speed and exchanges measured variable, setpoint, and start/stop commands and indications with the controller.

Adva Soft for Windows is a software application package that turns PCs into an open system for interfacing with ABB Master. The package includes several options (e.g., Adva Inform User API gives access to Advant Controllers database via relational/SQL–compliant database; SQL Net for TCP/IP provides access to Advant Controllers over TCP/IP networks [Ethernet, Token Bus, etc.]).

7.3 FISHER PROVOX

7.3.1 Introduction

Fisher[12] has developed a range of hardware/software packages to facilitate advanced control and to build a platform for the use of new techniques for better management of process plants. The severe competition and ever-increasing demand for cost reduction and environmental control will force companies to employ new methods for the control and operation of process plants.

Although it is possible to use third-party software application packages in any control system, it will be much easier if the system vendor offers these as standard or optional packages. This will also encourage application engineers to apply the available software packages to the control of their processes. The application packages, which are readily available in PROVOX, are briefly described as follows:

- Intelligent Sensor uses a neural network to measure difficult process variables such as quality.
- Intelligent Tuner automatically tunes the PID control loop. This facility is especially helpful during commissioning and startups. It is also useful for loops that have variable control parameters (i.e., their process gain and or time constants may change with flow, pressure, level, or temperature).
- Multivariable Fuzzy Controller is useful for noisy or nonlinear process control loops.
- SIMVOX simulation package is used for testing control strategies and training operators.
- ENSTRUCT provides a set of preengineered control solutions to help with the configuration of the control system.

PROVOX Management Packages and Interfaces are indicated as follows:

- Safety system interface: Pyramid Integrator/CHIP interface, to use dual or TMR safety systems, such as Allan-Bradley or Triconex.

- Computer/hiway interface package (CHIP): For efficient data exchange between PROVOX and third-party computers (DEC VAX, HP 1000 and 9000, IBM XT, AT, PS/2 and 7500 series).
- Batch data manager: Uses a relational database to manage batch processes.
- Expert system data server: Provides interface between PROVOX and G2 from GENSYM.
- Trace and Tune: For testing control loops by using simulation.
- ENSTRUCT: For the development of an application database by using a library of preengineered functions (algorithms), such as interlocking, sequence control, continuous control.
- SIMVOX: For testing and troubleshooting of PROVOX application packages and operator interface. It can also be used for operator training.
- Model prediction controller.
- Easy Tuner: For tuning of PID control loops in PCS.
- Neural networks

The application packages discussed previously are readily available in PROVOX and should be employed to improve the performance of PCS. It is the responsibility of application engineers to use them in control loops that do not function satisfactorily or need improvements to operate at optimum levels. In these competitive times, where the survival of businesses depends heavily on increasing profits and reducing costs by improving the quality and quantity of products, it is vital to employ new techniques, such as neural networks, optimization, and simulation to meet objectives.

7.3.2 PROVOX Communications

The PROVOX Communication facilities are designed to yield an open system architecture for easy integration into management systems. Three distinct levels of communications are provided:

- Local area network (LAN) for interface to business computers
- Process control network (PCN) for primary interface between control units and operator stations
- Field I/O bus for efficient interface between control units and I/O subsystems

The LAN employs Ethernet (IEEE 802.3), which is by far the most widely used communication standard. It is used by many systems for both business systems and process control systems. PROVOX LAN uses the computer/hiway interface package (CHIP) and network interface unit (NIU) to provide a gateway between computers and the PCS real-time database. Popular computers such as DEC VAX, HP 1000 and 9000 series, and IBM XT, AT, PS/2, and 7500 series are supported. Applications such as manufacturing scheduling,

146 | INDUSTRIAL PROCESS CONTROL

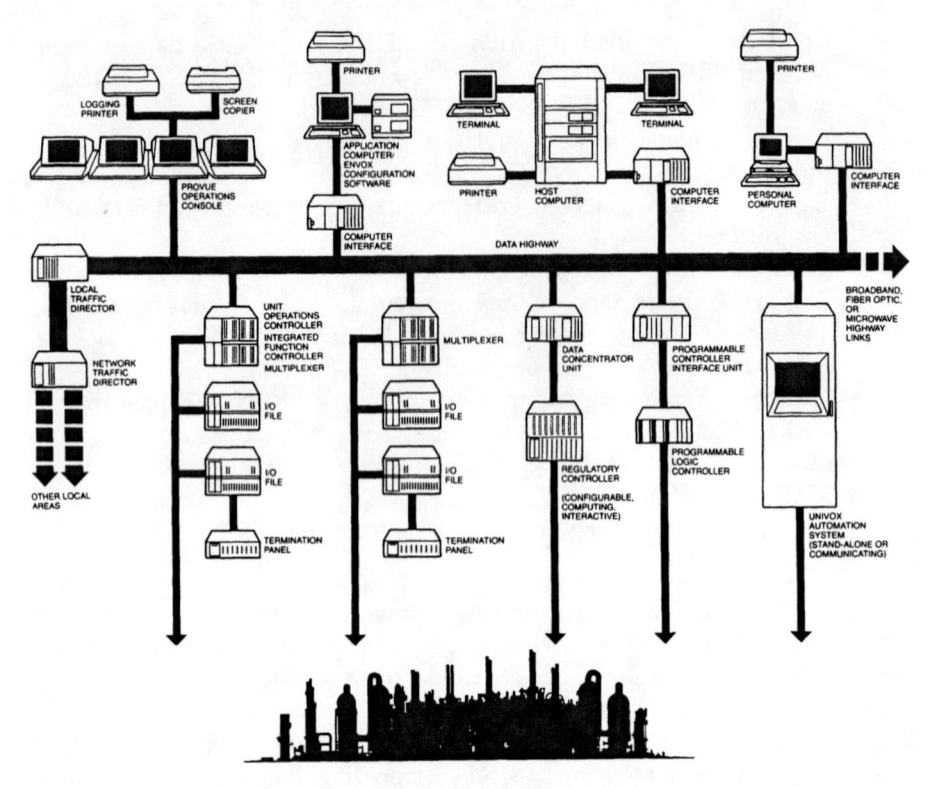

Figure 7-3 PROVOX system configuration and main components.

optimization, and simulation for training can be easily implemented in such computers.

The process control network (PCN) is the second level of communication and is the main carrier of data between the control units and the operator stations. It is based on the IEEE 802.4 standard (token passing bus) with a communication rate of 5 MBaud. The PCN can be extended to remotely installed subsystems by the use of fiber-optic extenders.

Field I/O communication is a token passing bus, which provides interface between control units and I/O modules. When I/O subsystems are installed remotely (e.g., near the process), a substantial saving in wiring and cabling efforts may be achieved. The I/O modules can be located up to 1,500 m away from the control unit. Figure 7-3 shows the system configuration and the main components of a PROVOX system.

7.3.3 PROVOX Operator Interface

PROVUE Operator Console is a user-friendly and flexible operator interface that uses an alarm data hierarchy system for efficient operation of process plants. The alarm hierarchy system allows for division of plants into several levels (e.g., plant management areas, process plant areas, operation areas, equipment/facilities areas).

The application window allows the monitoring of application packages, such as optimization and spreadsheets, which will be useful for quality control or decision making on alternative strategies. Various types of standard displays, such as overviews, graphics, trends, and averages, are available.

7.3.4 PROVOX Control Units

The PROVOX range of control units provides for all possible types of process control applications, such as continuous, logic, and batch control. The SR90 control unit can be configured for three different applications:

- Unit operations controller (UOC) for advanced batch control
- Integrated function controller (IFC) for continuous control
- Multiplexer (MUX) for data acquisition and a remote terminal unit (RTU)

The SR90 control units can be provided with redundancy in one-for-one to one-for-four units. The controller I/O facilities will satisfy the requirements of any type of PCS. The PROVOX control unit I/O subsystem has the following features:

- Remote I/O modules: the I/O cards can be located up to 1,500 m from the control unit.
- Flexibility: four types of I/O cards will cover all process I/O requirements.
- Smart transmitter interface card allows diagnostics and calibration from the central control room.
- Advanced automation solutions such as intelligent fuzzy logic, intelligent tuner, and intelligent sensor (uses neural networks).
- Redundancy: an I/O card can support up to eight cards.

7.3.5 PROVOX Software/Computing Facilities

By using the PROVOX CHIP software package, it is possible to integrate high-level/supervisory computers (from popular PCs or minicomputers such as DEC, IBM, HP) into PROVOX for implementation of some useful application packages. Application packages currently available are as follows:

- Data Historian: A modular structured system for collecting data, providing customized reports, trending, and performing statistical analysis for quality control
- Batch Data manager: Provides batch data management similar to that of the Data Historian. Event tracking and snapshot data gathering are additionally available

- Expert Systems: Interface to GENSYM G2 expert system for process data analysis and forecasting
- Third-party software packages

7.3.6 *PROVOX Engineering Tools*

The PROVOX database configuration package ENVOX uses a relational database and structural query language (SQL) for easy and efficient project engineering. The ENSTRUCT software package offers preengineering routines to help with application development, including the following:

- Discrete element control: for pump start/stop and interlocking
- Fixed sequence control: for process equipment startup and shutdown
- Single and multistream batch control

Graphics Tool Kit allows building and importing AutoCad graphics into PROVOX display system. SIMVOX allows troubleshooting control schemes and operator interface before the start of the plant. It can be used for operator training and control system validation by simulating process inputs/outputs.

7.4 HONEYWELL TDC 3000

7.4.1 Introduction

Since the introduction of Honeywell's TDC 2000[19] in the mid-1970s, numerous suppliers of instrumentation, control systems, and computers have introduced microprocessor-based DCS; however, most of these systems were unsuccessful, and quite a few disappeared from the market or are still struggling. Many companies found themselves on the verge of bankruptcy and were rescued by bigger brethren. What was the reason behind the TDC 2000's success during the first 5 to 10 years, where Honeywell had a lion's share of the market (more than 50 percent in some applications, like refineries, North Sea offshore)? Honeywell still sells more systems than others, and the total number of installed systems is more than 3,000; however, a few vendors, by using the latest hardware, software, and information technology ideas, and by offering superior systems, have made substantial progress and pose a serious challenge to Honeywell.

The main factors that contributed to the stupendous success of TDC 2000 were as follows:

- Simplicity
- True distributed control
- Robust architecture
- Reliable hardware and software
- Well-documented system

Many early control systems, which failed to satisfy the aforementioned requirements, were squeezed out of the market. Some vendors were promoting single-loop integrity, or even sold analog controllers as an integral part of their DCS. TDC 2000 systems, which were installed in the late 1970s and 1980s, are still in place, although upgrading of some units (e.g., basic controller, operator station) is inevitable; however, many so-called distributed control systems by other vendors had to be revamped and replaced by true DCS systems, normally by a different vendor.

Although TDC 3000 is far more powerful than TDC 2000, many useful features of the latter are retained in TDC 3000. Honeywell often offers the TDC 2000 components (e.g., basic controller, multifunction controller, data hiway) as an integral part of TDC 3000 systems; however, TDC 2000 components do not comply with open system requirements and do not yield a true integrated system configuration. Figure 7-4 depicts a typical TDC 3000 configuration. In the following six sections, a brief description of the main elements of TDC 3000 will be presented.

7.4.2 TDC 3000 Communications

TDC 3000 communication facilities include plant information network (PIN), universal control network (UCN), local control network (LCN), and Data Hiway (see Figure 7-4). PIN, UCN, and LCN conform to OSI standards. This

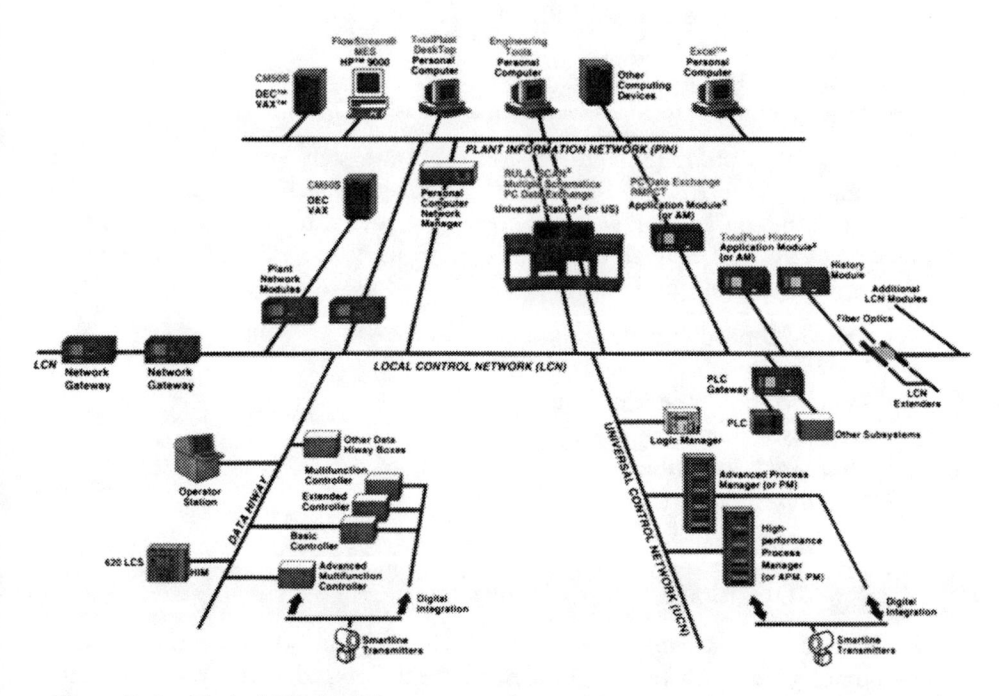

Figure 7-4 Typical TDC 3000 system configuration.

allows easy and efficient interfacing to foreign systems. The UCN is a high-performance, real-time, map-based network, which provides a secured multiuser communication network. The UCN is based on IEEE 802.4 (ISO 8802/4) standards and operating at 5 MBaud. It supports up to 32 devices (e.g., PM, LM) and provides peer-to-peer, multicast (server distributions), and broadcast (all devices process the message) communication modes.

The PIN facilitates data exchange between several LCNs. PIN is a broadband network and complies with the IEEE 802.3 standard. Coaxial or fiber-optic cables can be employed. Data transmission rate is 10 MBaud.

The LCN is the backbone of TDC 3000. It uses a proprietary token-passing protocol and includes several levels of error checking. It operates at 5 MBaud and is based on the IEEE 802.4 standard. Fiber-optic and LCN extenders can be used to cover large plant areas (e.g., where several outstations require control and monitoring from a central control room).

Data Hiway, the original TDC 2000 Hiway, operates at 250 kbits/second and has been enhanced to support the following devices:

- Hiway gateway for interface to the LCN
- Operator station (OS)
- Basic controller (BC)
- Extended controller (EC)
- Multifunction controller (MC)
- Advanced multifunction controller (AMC)
- Process interface units (HLPIU, LLPIU, LEPIU)
- Logic controller (LC)
- Triconex triple modular redundant controller (TMR)
- Miscellaneous interfaces (serial links, PCs)

MASLINK is the manufacturing automation system (MAS) network that accommodates supervisory controllers (MAS/C) for the management of such tasks as finishing, packaging, shipping, receiving, and warehousing. The combination of MAS/C and TDC 3000 process control systems yields a flexible manufacturing environment, and will provide for the following:

- Improved productivity
- Improved quality
- Increased equipment utilization
- Decreased operating costs
- Decreased dead times
- Minimized downtime

7.4.3 TDC 3000 Operator Interface

Universal station (US) is the primary TDC 3000 person–machine interface that communicates with other modules on the LCN and with process control units on UCN and Hiway. The US is a well-designed person–machine interface

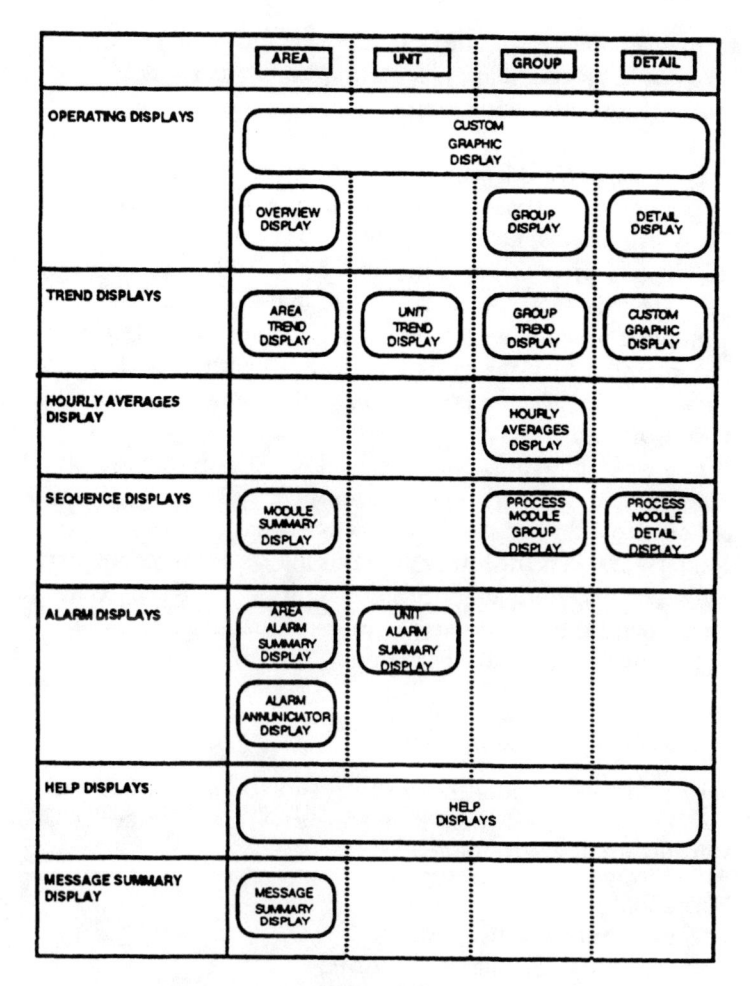

	AREA	UNIT	GROUP	DETAIL
OPERATING DISPLAYS	\multicolumn{4}{c}{CUSTOM GRAPHIC DISPLAY}
	OVERVIEW DISPLAY		GROUP DISPLAY	DETAIL DISPLAY
TREND DISPLAYS	AREA TREND DISPLAY	UNIT TREND DISPLAY	GROUP TREND DISPLAY	CUSTOM GRAPHIC DISPLAY
HOURLY AVERAGES DISPLAY			HOURLY AVERAGES DISPLAY	
SEQUENCE DISPLAYS	MODULE SUMMARY DISPLAY		PROCESS MODULE GROUP DISPLAY	PROCESS MODULE DETAIL DISPLAY
ALARM DISPLAYS	AREA ALARM SUMMARY DISPLAY / ALARM ANNUNCIATOR DISPLAY	UNIT ALARM SUMMARY DISPLAY		
HELP DISPLAYS	\multicolumn{4}{c}{HELP DISPLAYS}
MESSAGE SUMMARY DISPLAY	MESSAGE SUMMARY DISPLAY			

Figure 7–5 Universal station available displays.

that provides a comprehensive set of graphical/tabular displays for operation, engineering, and maintenance. Figure 7–5 shows the range of displays available to US users.

The US is provided with a security system to prevent unauthorized data access or inadvertent data entry. The security keyswitch allows three levels of access as follows:

- Operator level for normal plant operation (e.g., opening/closing valves, changing control setpoints, starting/stopping pumps)
- Supervisor level for altering plantwide parameters, such as flowrates for some products
- Engineer level to allow changes to configuration parameters (e.g., ranges, I/O addresses, control schemes)

A fourth level of access to allow monitoring only can be configured. Whenever operators or engineers make data entry, the system checks for the validity of the data and keystrokes. For example, if an operator changes a controller setpoint, the system will accept the data, if the control mode is right (i.e., auto mode). If it is in cascade mode, the system will reject the data entry and alert the operator by sounding an alarm or displaying an appropriate error message.

The standard US displays are comprehensive and will satisfy the requirements of all types of processes and plants. Trend displays can show multitrends (up to 24) for a large process area. Various time bases are available. Trends may be used to monitor process operations or for PID controller tuning.

The US alarm system design is based on a hierarchical and logical structure. Alarm priority, displays, reports, keys, and messages are critical for the efficient operation of large plants and plants with complex processes. Alarms and warnings are invaluable tools for operators to prevent costly shutdowns or damage to equipment and personnel injuries.

In addition to US, TDC 3000 offers operator station (OS) and universal work station (UWS). OS is the hiway operator interface. For small plants, a hiway with hiway-based control/data acquisition units (e.g., basic controller, multifunction controller, HLPIU) and an OS may be adequate. The UWS is used for engineering, maintenance, and supervision.

7.4.4 TDC 3000 Control Units

Honeywell offer several different control and data acquisition units to fulfill various process control and logic requirements. They include TDC 2000 and TDC 3000 control units as follows:

- Basic controller (BC)
- Extended controller (EC, AEC)
- Multifunction controller (MC, AMC)
- Process interface units (HLPIU, LLPIU, LEPIU)
- Logic controller (LC)
- Critical process controller (e.g., Tricomex TMR)
- Logic manager (LM)
- Process manager (PM, APM)

In this section, we will study the process manager only; readers may obtain technical documents from Honeywell for other control units. The advanced process manager (APM) and process manager (PM) offer highly flexible I/O facilities to satisfy the requirements of various processes. Remote I/O can be handled by the use of fiber-optic. This approach will result in substantial savings on cabling, wiring, cable trays, and the like. Various types of I/O, including high-resolution pulse inputs (20 kHz) and digital inputs with 1 msec sequence of event (SOE) facility are supported.

In addition to a comprehensive library of control algorithms, APM is equipped with a PCS-oriented programming language called CL/APM. CL/APM can be used to write programs for batch control, hybrid applications, and specialized computations. The APM database, including other LAN-based control units, is shared and available to CL/APM programs. Many types of data (e.g., arrays, flags, timers, and strings) can be configured into control strategies.

For critical applications, where shutdowns are costly (e.g., process shutdown, emergency shutdown, unstaffed satellite platforms) either a dual PM or a critical controller may be used. The former not only reduces the hardware/software variety but also avoids the use of slow serial interfaces. Some users may not be willing to employ APM for ESD, in which case I would suggest using a solid-state system for the highest levels of shutdown (e.g., SIL 4, SIL 3 and 2) (APS, ESD1, ESD2) and a dual APM for the other levels (SIL 1 and 0). SIL 3 and 2 normally have less than 100 I/O, and the use of a dual solid-state system will be acceptable. PM and APM are provided with some useful application packages; these are studied in Section 7.4.7.

7.4.5 TDC 3000 Host Computer

Two LCN-based units, application module (AM) and history module (HM), provide for functions that are normally in the domain of host and supervisory computers. Additionally, other computers or PCs can be easily interfaced to LCN via computer gateway or PC gateway. Application and history modules have access to all UCN and hiway device databases (PM, BC, EC, PIU, etc.). This includes both process data and configuration parameters.

The HM stores process historical data, displays and graphics, configuration database, and system records in permanent memory. The HM gathers information automatically and without need for operator intervention. Historical data includes averages and samples of process variables, process events, operator actions, system hardware events, and diagnostics. The HM hard disk can be provided with backup to enhance the reliability of the system.

The AM is used for plantwide strategies, which are normally outside the scope of a process manager or other control units. Such tasks as simulation, optimization, and multiunit control strategies can be implemented in the AM. A comprehensive library of algorithms is readily available for use by the application engineer. Additionally, the AM offers a PCS-oriented programming language, control language (CL), which can be used to write custom control schemes such as startups, shutdowns, batch control, and multivariable control.

For systems requiring interface to business and management computers, the TDC 3000 computer gateway allows integration of the control system and DEC computers. Applications such as materials and cost control, inventory, spares control, job scheduling, modeling, and global optimization schemes require mainframes or minicomputers. The TDC 3000 CM505 provides a secure link between VAX computers and various control modules databases. Programs in FORTRAN or PASCAL have easy access to LCN units database by the use of standard subroutines. These programs are fully integrated with other LCN functions.

7.4.6 TDC 3000 External Interfaces

The TDC 3000 is equipped with a range of gateways and interface facilities for efficient communication to foreign systems. A summary of such interface devices is given as follows:

- Computer gateway: For communication between LCN and foreign computers
- PLC gateway: For communication between LCN and Modicon or Allan Bradley PLCs
- PC gateway: For interface between LCN and IBM-compatible PCs
- Critical process controller interface: For interface between hiway and Triconex TMR PLCs
- Process manager serial interface: Provides RS 232 serial link for communication with other systems (via MODBUS Protocol)
- Data hiway port: For interface between hiway and foreign PLCs
- Plant network module: For interface between LCN and DEC or other computers via Ethernet

7.4.7 TDC 3000 Application Packages

The application packages can be implemented in AM, APM, LM, AMC, or CM. Some application packages may require a dedicated PC, in which case the PC can be interfaced to the TDC 3000 via a PC gateway. End-users, in addition to a host of advanced control strategies offered by TDC 3000, can use the control language to develop their own specific advanced control schemes. For example, control language together with the standard AM algorithms can be used to develop simulation models, optimization models (e.g., hill climbing), batch control, and multivariable control. A list of some of the available application packages is as follows:

- Loop tune
- Horizon predictive control
- Real-time statistical process and quality control
- Digester control (batch and continuous)
- Turbine generator management
- Tank farming management
- Modular batch automation

7.5 SIEMENS SIMATIC PCS7

7.5.1 Introduction

One of the largest North Sea offshore oil and gas production plants (Oseberg) employed Siemens Teleperm M and Simatic PLCs for the control and safety systems. We endeavored to reduce the hardware and software variety by insisting that all major packages should use SIMATIC PLC for control, rather than using proprietary control systems. Only one major package vendor (for generators and compressors) and a couple of small vendors (metering) insisted on using their own control systems.

Within the control and safety system, distributed supervisory, control and safety (DISCOS) system, hardware variety was reduced by using one type of I/O card for control and safety (PCS, PSD, ESD, F&G) systems. To reduce communication problems (interface between DISCOS and other systems, i.e., generators, compressors, and metering) and standardize on serial links, we insisted that those packages that use proprietary control systems should implement the DISCOS communication protocol.

It would appear that DISCOS was the first control and safety system where the notion of integrated system configuration was introduced. I worked on the project as a systems engineer from early 1986 to late 1988 and gained considerable experience in system interfacing. Siemens benefited from providing a very large system (more than $50 million) and also realizing the potential of integrated systems. As a result, the SIMATIC PCS7 is one of the most advanced control and safety systems, and it complies with OSI standards. Virtually any application (e.g., antisurge control, fiscal metering, and batch control) can be easily implemented. I must emphasize that even after more than 10 years since we introduced the integrated system architecture for Oseberg, most control and safety system vendors do not offer integrated systems.

7.5.2 SIMATIC PCS7 System Overview

The SIMATIC PCS7 system[20] offers a modular and scalable distributed system configuration that can be applied to process plants of virtually any size. Well-designed communication networks and powerful operator stations ensure that information will reach wherever and whenever it is needed. The SIMATIC PCS7 database uses popular software/IT packages and yields a highly reliable and flexible relational database. The system employs the Windows NT/UNIX operation system, X-windows/Motif graphical user interface, SQL database, Oracle, and event notification mechanism. Figure 7–6 shows a typical SIMATIC PCS7 system configuration.

The main components of the system are:

- System communications
- Operator interface
- Control units
- Supervisory computer
- Engineering interface

In the following sections, the major components of the SIMATIC PCS7 system will be studied.

156 | INDUSTRIAL PROCESS CONTROL

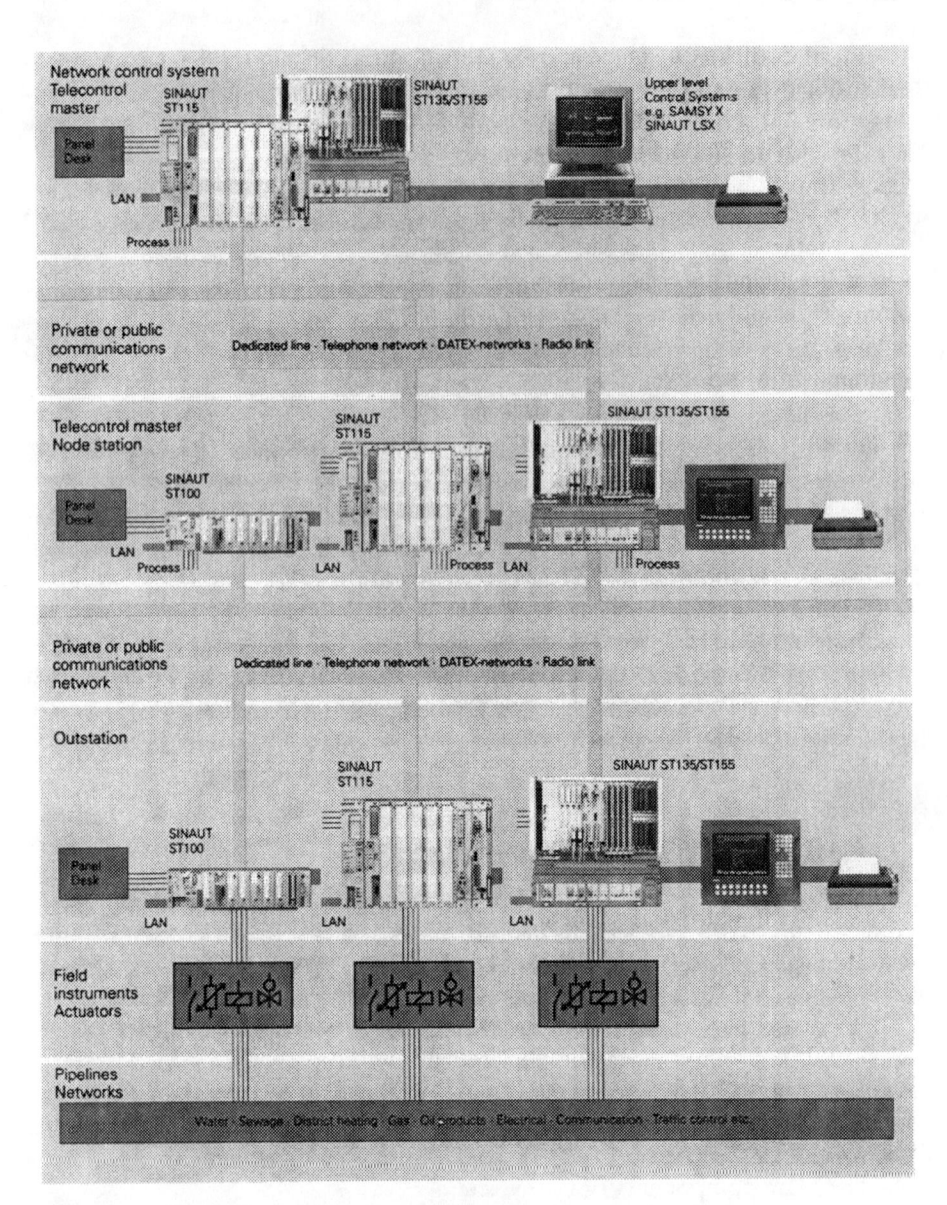

Figure 7–6   SIMATIC PCS7 system configuration.

7.5.3 SIMATIC PCS7 System Communications

The SINEC H1 network is the backbone of SIMATIC PCS7. It is the seven-layer Ethernet, designed in compliance with the IEEE 802.3 Standard and CSMA/CD data control over coaxial and/or fiber-optic media. The system executes multiple protocols simultaneously on SINEC H1 in order to ensure the high performance required for real-time process control systems and the open system interconnect necessary for interface to external (business-type) systems. The SINEC protocol provides a communication interface similar to that of MAP 3.0, which allows easy interface between SIMATIC PCS and those systems implementing MMS functions of MAP 3.0. The various SINEC protocols are STF (equivalent to the MMS functions of MAP 3.0), TCP/IP (for exchange of files between SIMATIC PCS operator stations and to other systems), and NSF (for presentation of historical data to multiple nodes).

The system employs "demand scan" and "report by exception" methods for exchange of information between various nodes. The result is a high-performance, low-traffic communication network. This is in contrast with systems using polling method, where a large portion of communicated data is unnecessary. Another useful feature of SINEC H1 is its ability to use both media (primary and backup) in parallel during normal operations to distribute tasks, which minimizes traffic load. In this mode, two communication protocols can use the network simultaneously. Figure 7-7 shows the SINEC H1 architecture and various nodes/applications. The applications may be provided by Siemens or other vendors, provided they conform to the open system interconnect standards.

7.5.4 SIMATIC PCS7 Operator Interface

The SIMATIC PCS7 operator station provides all of the necessary displays and functions for control and monitoring of process plants. Up to 16 operation nodes can be supported by a SIMATIC PCS7 System. The system utilizes the Windows NT/UNIX operating system and X-Windows/OSF Motif graphical user interface (GUI). This yields a highly user-friendly environment for operator and engineer interface. The system offers multiapplication, client-server support, where data from several servers can be displayed on an operator display.

Operator station employs a flexible paging hierarchy, where each display may have one or more associated displays. For example, a graphic display may have associated trends and help displays.

In addition to standard displays (e.g., free-format dynamic graphics, groups, details, trends, alarms), action requests are available to help operators during process upsets and emergencies. Action request displays are linked to process events (e.g., pressure high- or low-level alarms) and can provide useful instructions (preconfigured) to remedy process upsets. They can provide information (e.g., the degree of severity of the process upsets), prompt the operator for specific actions, check process equipment, or change control parameters.

7.5.5 SIMATIC PCS7 Control Units

The SIMATIC control units are designed to conform to the latest IT standards and meet the requirements of virtually all control functions. They can be applied to processes that need very fast sampling, multitasking/parallel processing, or special algorithms and calculations. SIMATIC S7-400 range of controllers

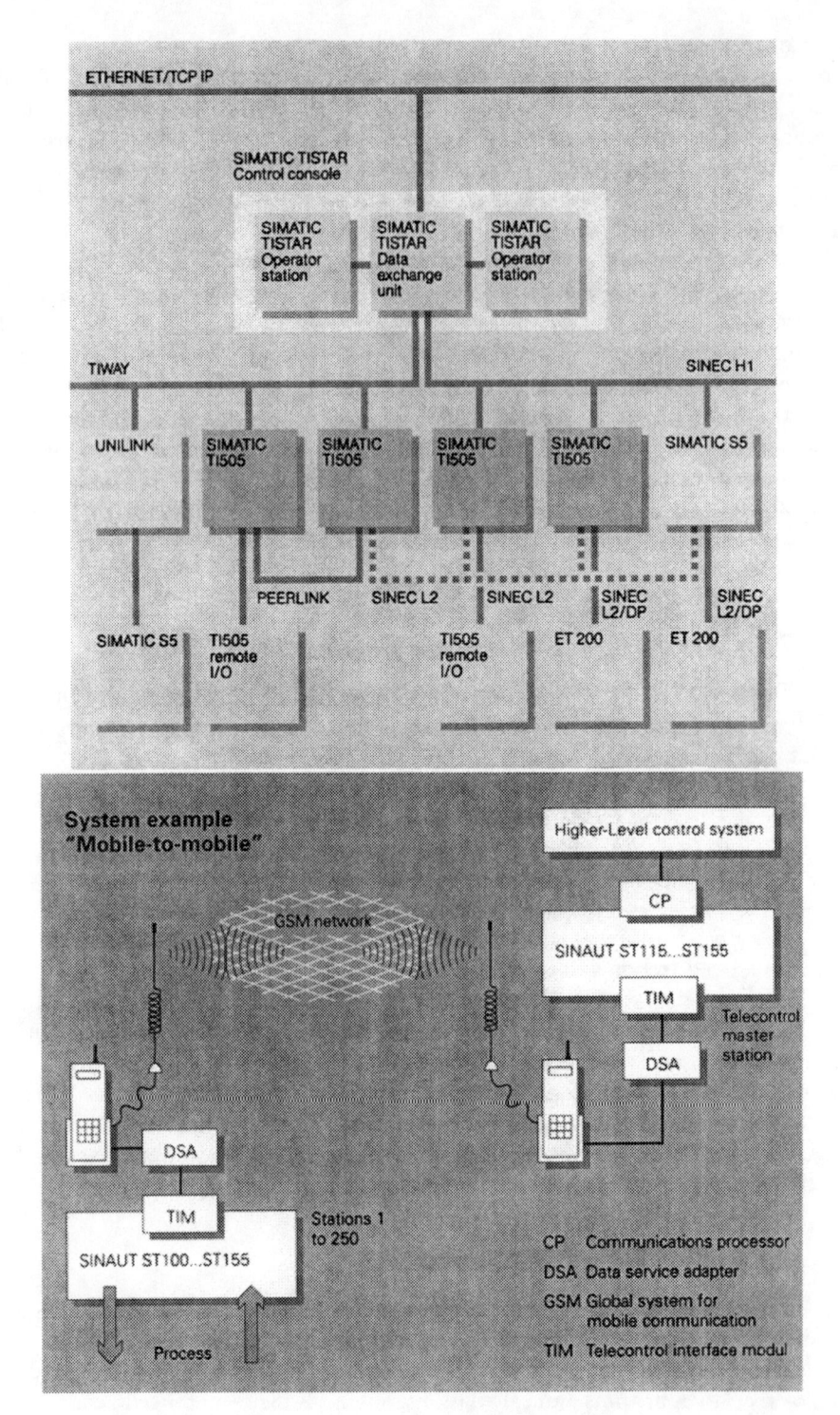

Figure 7–7 SIMATIC PCS7 system communication architecture.

have been used in applications that are normally in the domain of specially designed/proprietary systems. Following are some examples where SIMATIC controllers have been used:

- Batch control
- Continuous control
- Safety critical, ESD, F&G, atomic power plants (TUV Class 8)
- Fiscal metering
- Antisurge control
- Machine condition monitoring
- Manufacturing
- Subsea master control
- Boiler

SIMATIC controllers can be configured with up to four CPUs, 8 Mbytes RAM, handle more than 10,000 I/O, and process PID loops faster than 10 msecond. Many equipment manufacturers employ SIMATIC controllers as their standard hardware to control and monitor their packages. SIMATIC controllers use the UNIX operating system where software packages, written in UNIX by other vendors, can be easily implemented in SIMATIC PCS.

7.5.6 SIMATIC Supervisory Computer

The supervisory computer provides process management services such as display building, historical trending of process parameters, archiving of data, alarm/event/data reporting, and the batch control language (BCL) for advanced batch recipe control and dynamic scheduling.

7.5.7 SIMATIC Engineering Interface

The SIMATIC PCS7 controller database is developed by using the application productivity tool (APT). The system relational database and control/logic schemes are configured graphically. SIMATIC APT is an object-oriented tool, which is of great help to process/control engineers who think of a plant as a combination of objects (process equipment). Once an object (e.g., a pump) is configured, it can be used as many times as required because each one will have a unique tag number.

Sequential control strategies are built using standard APT sequential function charts (SFCs). SFCs represent the process state and transition diagrams. They show process phases and steps graphically. For regulatory control, APT offers continuous function charts (CFCs). Function blocks (FBs) are used and shown graphically to build control schemes, such as cascade, auctioneering, ratio, and so on. Siemens also provides a telephone service, where online fault analysis and maintenance routines can be accommodated.

7.6 SILVERTECH SENTROL

7.6.1 An Overview of Sentrol

Silvertech[21] offers a host of subsystems under the Sentrol name, which will satisfy the control, safety, and management requirements of small to medium-sized process plants. The Sentrol main subsystems are indicated as follows:

- Sentrol 1000: Supervisory and Display
- Sentrol 2000: SCADA
- Sentrol 3000: Industrial Control
- Sentrol 4000: PCS
- Sentrol 5000: ESD
- Sentrol 6000: F&G

These subsystems can be used either individually or as an integrated control and safety system. If used as an integrated system, the unified control system (UCS) offers a unified hardware/software architecture for the entire plant's control, safety, and monitoring system. Orthodox control and safety systems employ a variety of hardware/software and slow serial links for PCS, ESD, F&G, and package control and their interfaces.

Silvertech provides total plant automation solutions. They employ the latest hardware, software, and IT methods and comply with IEC 61508 and other emerging standards. The system hardware can be independent and is generally based on GE Fanuc PLCs and the Genius I/O communication system. Figure 7–8 shows the Sentrol system configuration.

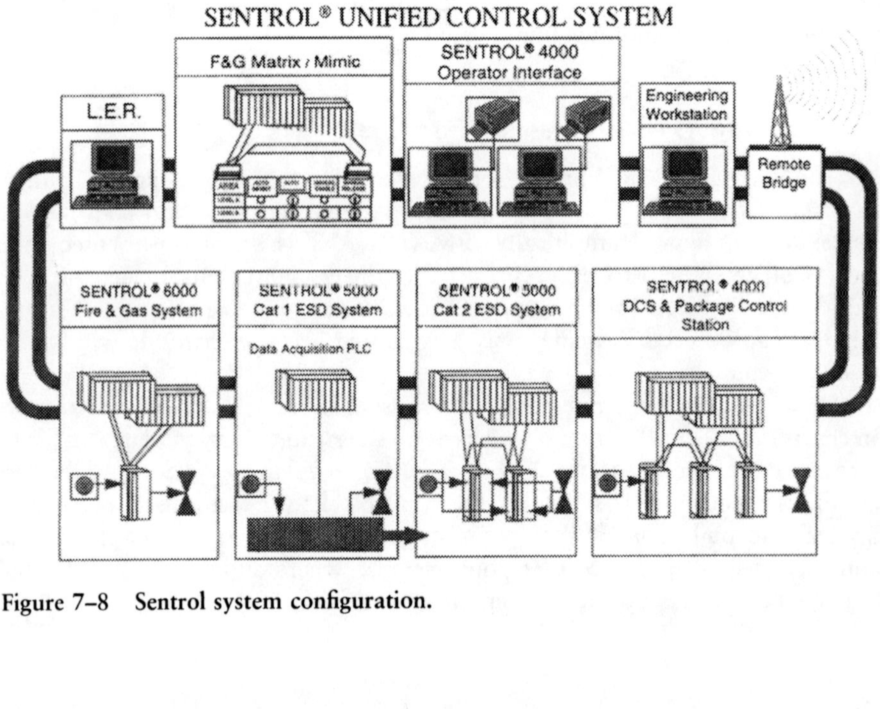

Figure 7–8 Sentrol system configuration.

7.6.2 Sentrol Internal Communications

Sentrol employs several communications media, including Ethernet LAN (TCP/IP, MMS, MAP 3.0), Genius LAN, CCM2 (MODBUS protocol via RS 232/422) and SNP (RS 485). Figure 7–8 shows various available communication facilities.

Ethernet LAN (IEEE 802.3) employs thick wire and thin wire coaxial cables with a transmission rate of 10 MBaud. The console controllers are responsible for the management of data exchange between the operator station and the remote devices. The data exchange between control units and operator stations is handled on a polled or an event-driven basis. During normal operations, data packages are sent cyclically, but when an event (alarm, status change) occurs, the package containing the event is transmitted instantly. The communication unit scan is typically less than 200 msecond.

The Genius LAN is a GE Fanuc proprietary network, which uses a high-integrity deterministic protocol. It features token passing with fast recovery, peer-to-peer and broadcast communications, automatic acknowledgment of the transmitted data, and 2oo3 dipulse data bit voting. The Genius LAN data transfer rate is at 153.6 kBaud, 76.8 kBaud, or 38.4 kBaud for cable lengths of 1 km, 1.5 km, or 2 km, respectively. The maximum number of nodes on the LAN is 32. The medium is a single twisted pair. Broader areas can be covered, if repeaters and/or fiber-optic are used. Distances longer than 10 km can be handled. The bus is generally used for communication to the Genius I/O and for intersystem interface for control data.

7.6.3 Sentrol Operator Interface

The Sentrol operator interface uses a dual console controller, with typically 1 to 12 VDUs, along with message/alarm printers and color printers. The console controller is based on IBM/PC-compatible industrial computers. All operator stations have full access to all of the system functions and printers.

The console controllers operate in a hot standby mode. The active controller manages the exchange of data with the remote control units and continually synchronizes the standby controller. In the event of the active controller failure, the standby resumes control automatically and without loss of data. Process data from the remote control units is received on an event-driven basis. The operator interface employs a modular design philosophy and yields a flexible system upgrade path. It reduces the spare requirements and maintenance costs considerably.

The redundant console controller provides a large database and can be used for supervisory/management functions. The system is Windows NT or UNIX based and includes extensive graphics and processing capabilities.

7.6.4 Sentrol Control Units

The Sentrol control unit (CU) normally employs GE Fanuc Series 90 PLCs and Genius I/O hardware. The CU is equipped with a full range of interfaces, including process I/O (analog, digital, RTD, etc.), power supplies (AC, DC), LAN, and a battery-backed real-time clock.

The CUs are available in different sizes, comprising multiples of five-slot and/or nine-slot racks. The CU central processor unit (CPU) supports a comprehensive range of instruction sets, hard and soft interrupts, four levels of password protection, several levels of fault analysis routes, and watchdog facility (software plus a hardware backup).

The CU supports a local bus, called Local Genius Hiway. If it is necessary for two or more CUs or remote I/O units talk to each other, this configuration is very useful.

A superior feature of the Sentrol CU is its application in safety critical systems. Simplex, duplex, or triplex configuration can be used for process control and safety (PSD, ESD, HIPPS, F&G) systems. This unified hardware reduces the burden of spare parts management and maintenance effort and eliminates the need for slow and troublesome serial links. The flow of information between safety system and operator stations/host is substantially better in unified and integrated systems than orthodox systems (i.e., systems that employ serial links between PCS and safety subsystems). Figure 7–9 shows a triplex ESD system configuration.

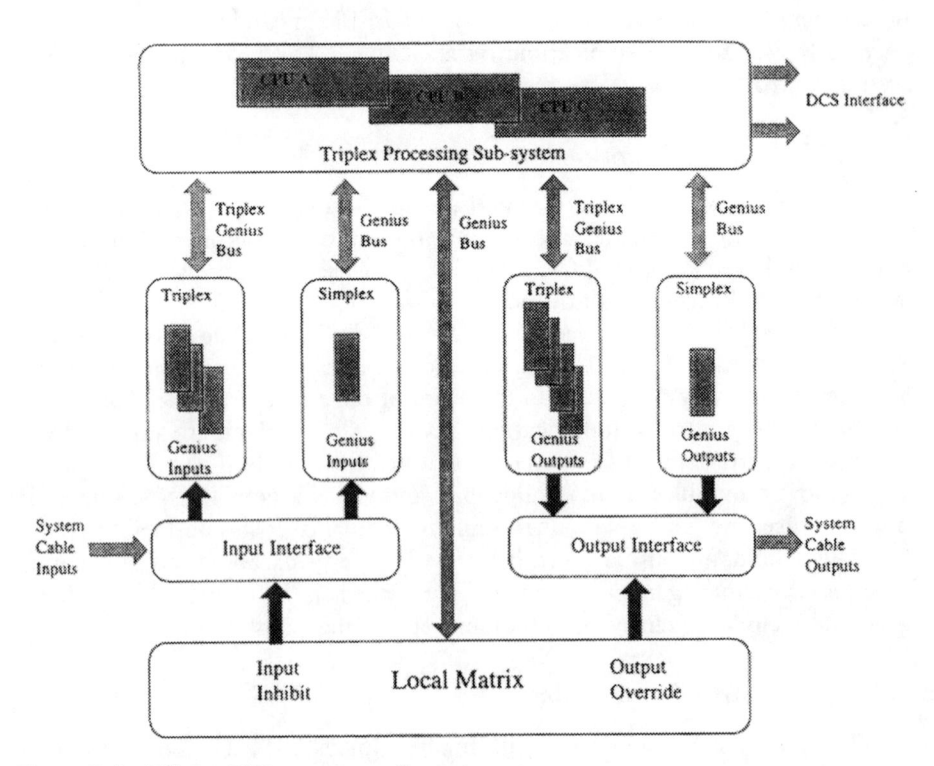

Figure 7–9 Triplex ESD system configuration.

The Genius I/O system allows the distribution of I/O subsystems over a large area. This will reduce the wiring and cabling from field transmitters and valves to the PCS significantly. It also allows CU-to-CU communication, which is useful for multivariable applications, where variables may come from different process areas.

7.6.5 Hazardous Area Applications

Silvertech has developed a range of remote terminal units (RTUs) that were specifically designed to operate in Zone 2 hazardous areas with Zone 1 areas interfaced to the system via IS barriers. Each of the RTUs is certified by an independent third party to Ex n rating and contains remote I/O modules or PLCs making Sentrol RTUs ideal for applications such as wellhead platforms, drilling rigs, offshore platforms, or FPSO projects. RTU-to-MTU communications are via radio, satellite, modem, or high-integrity databus, where large savings can be achieved because of the significantly reduced field cabling requirements.

7.6.6 Powertools

Silvertech has developed a formal system, Powertools, that allows a computer-based definition of the I/O and logical processing requirements for large-scale control and safety systems. The logical processing requirements are referenced to a set of predefined logic function blocks. Function blocks define a particular control function and may be used repeatedly to ensure a consistent approach to the implementation of that particular control function. The system has the following benefits:

- Automatic validation of entered data against predefined rules
- Automatic generation of system documentation, including drawings, wiring schedules, and test schedules
- Automatic generation of PLC software (ladder code), supervisory and display system database and system configuration data
- Controlled and efficient management of change
- Reduced life cycle costs
- Enhanced quality assurance

This approach reduces the effort required to perform the repetitive tasks needed to complete the detailed engineering for a system and thus allows the emphasis to be placed on validation of the base design. Because the control and safety systems market is increasingly focusing on system validation, Powertools is being established as the frontrunner of the next generation of control and safety system specification and implementation tools. Powertools offers significant technical and commercial advantages, including savings achieved by the early identification of potential design errors, allowing subsequent modifications to be incorporated with maximum efficiency.

The Powertools system is applicable to all types of systems, including DCS, ESD, F&G, HVAC, and any PLC-based control systems. Powertools is also applicable to all types of hardware platforms, including PLCs from GE Fanuc, Allen Bradley, Modicon, Siemens, and so on.

7.7 SIMRAD AIM

7.7.1 An Overview of AIM

Simrad's AIM control system[8] was developed from the existing technology used for vessel management and dynamic positioning and introduced to the process industry in the late 1980s. Simrad's proximity to the Norwegian offshore installations and close link with computer-aided design, analysis, and simulation (CADAS) are driving forces behind the success of the system and its acceptability by major Norwegian end-users.

AIM is one of the most advanced control and safety systems available in the marketplace and offers the following features:

- A true open system architecture
- An integrated system configuration (PCS, ESD, F&G, DPS, VMS)
- Efficient database configuration and documentation
- Availability of CADAS[7]

AIM has been installed in several large offshore oil/gas production installations with a large number of inputs/outputs. As the use of floating vessels (instead of platforms) for offshore production is becoming popular, and Simrad offers a true integrated architecture and many years of experience in vessel automation and dynamic positioning, I have no doubt that AIM will become a popular system, especially in the Norwegian sector of the North Sea. Figure 7–10 shows a typical AIM system configuration.

7.7.2 AIM System Internal Communications

AIM employs Ethernet with TCP/IP protocol for the main system LAN. This yields an open communication system, which facilitates easy interface to other systems using Ethernet and TCP/IP standard. Where distributed I/O subsystems are required, an I/O bus is available, which will provide interface between control units (process stations) and I/O units.

If the foreign system interfacing AIM does not support Ethernet, the two systems will be connected via a serial link. AIM supports some popular serial links (e.g., MODBUS via RS 485). Figure 7–10 depicts the AIM system configuration and the communication facilities.

7.7.3 AIM Operator Interface

The AIM operator stations employ UNIX rather than a proprietary operating system. This fact, together with the use of Ethernet for communications,

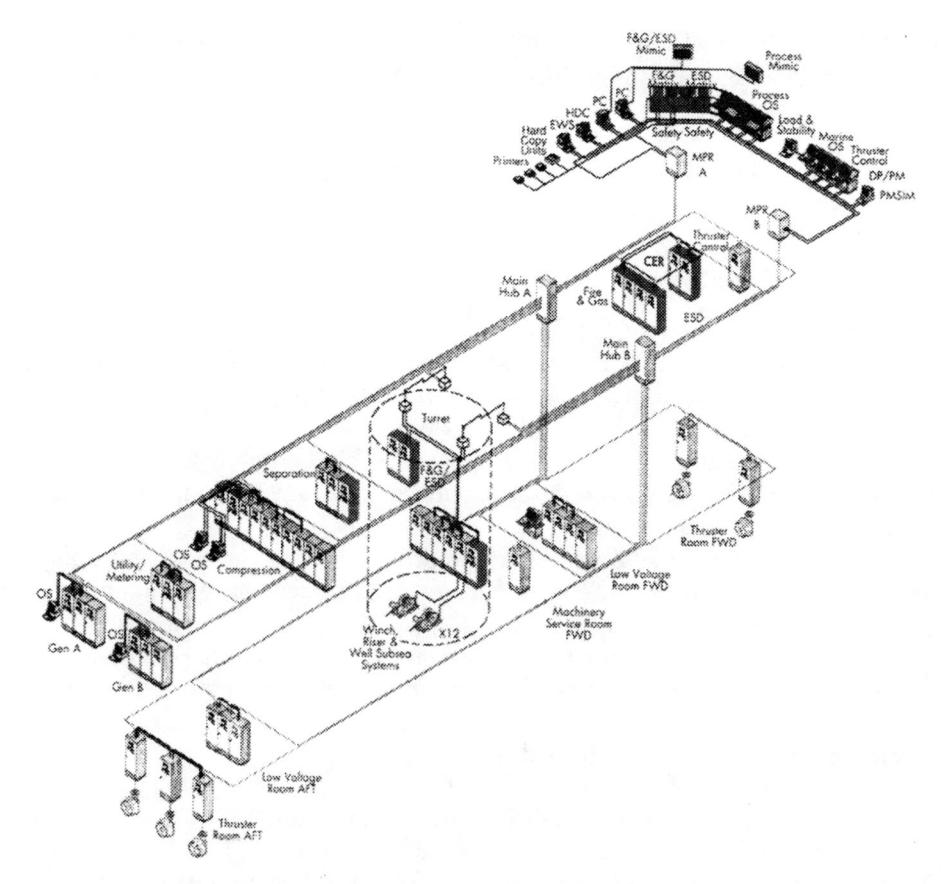

Figure 7–10 Typical AIM system configuration.

creates a true open system environment. It is possible for AIM end-users to procure software/hardware packages for the system from other vendors (provided they comply with ISO standards) and use them with minimum effort.

AIM Operator Stations provide all of the standard displays necessary for efficient monitoring and control of a plant, as follows:

- Overview
- Graphic (process flowcharts)
- Details
- Trends
- Diagnostics
- Alarms

The main difference between AIM and other PCS operator stations is that the former stores the graphic display information in the control unit's memory, rather than in operator stations. The advantage in storing the graphic database in control units is that once the control unit database is revised, all of the relevant operator station graphics will be updated automatically. This eliminates the need for updating several operator stations' memory or multiple displays. For example, if a signal's description is revised, several displays, which show the signal, will have to be revised. And if the system employs several operator stations, each station database has to be modified (in other PCS operator stations).

7.7.4 AIM Control Unit

The AIM control unit, called process station (PS), is one of the most advanced control units in the marketplace. Because it uses the UNIX operating system, it offers an open platform for import of third-party software packages. At present, UNIX is the most popular real-time multitasking operating system, which implies that numerous software houses are familiar with UNIX and write their control software (e.g., simulation, optimization, fuzzy logic, and neural networks) in a UNIX environment.

Process station, in addition to storing process related data (e.g., set points, ranges, control parameters), holds monitoring/graphic related information, too. Such data as signal tag and description, graphic lines, symbols, and connections are also stored in the PS memory. This is an interesting feature that prevents confusion and reduces the effort required to update a multiple-operator station database. This function is especially useful where revision of a database may affect several multilayer graphics/mimics displays.

PS has been applied to PCS, ESD, F&G, vessel control (VC), ballasting control (BC), and thruster control (TC). For safety systems, TUV Class 6 is available. Where floating vessels are used instead of platforms, AIM will offer a true integrated system. There are very few process control systems that have been applied to the control and monitoring of floating vessels.

Another area where PS offers support is simulation. Simrad, being a member of the CADAS project, offers a simulation package that not only helps with the design of control systems and operator training but can also be used in the design of process systems. This subject is covered in detail in Section 7.7.7.

7.7.5 AIM Host Computer

Because AIM uses OSI standards for the LAN and UNIX for operator station and control unit, it is possible to easily interface hosts of any size (PC, mini) and manufacture to the system.

7.7.6 AIM External Interfaces

The AIM control unit provides interface to foreign systems via serial links (e.g., RS 232, MODBUS, Siemens protocol); however, if foreign systems offer Ethernet, it is more efficient to interface the two systems via Ethernet rather than serial links. The process station has been applied to the control of various packages (e.g., generators, compressors, switchboards). This will eliminate the need for serial links.

7.7.7 AIM Application Packages

The process station is provided with special function modules for the control and monitoring of various process facilities (e.g., valve, pump, motor). For highly nonlinear systems, where a PID algorithm does not function satisfactorily, AIM offers a fuzzy logic controller. For control loops, which require variable PID parameters, an auto-tuner algorithm is provided.

Simrad, in conjunction with leading Norwegian companies (Aker, Kvaerner, Norsk Hydro, SINTEF, and Statoil), has created a life cycle simulation package called CADAS. CADAS is a tool for integrated process design and uses dynamic simulation, where process models and control schemes run concurrently. By employing CADAS at an early stage of a project, perhaps during conceptual studies, it is possible to optimize process systems/equipment and their control, before it is too late to carry out major revisions. The following benefits can be realized by using CADAS in the early phases of a major project:

- Knowledge buildup
- Closer cooperation between process and instrument teams
- Evaluation of alternative systems
- Process optimization and control by analyzing startups, shutdowns, control strategies, operability
- Verification of control logic and alarms
- Efficient operator training
- Improved quality
- Cost reduction in design, engineering, testing, revisions, training

Early simulation and analysis, combined with top-down design methodology, is an intelligent way of designing process systems. This supports the natural iterative process design. Iteration is normally used in operational research techniques to compute optimum solutions. Figure 7–11 shows how cost savings are possible by using CADAS in the early stages of a project.

The first step in implementing CADAS is to use process flowsheets to build graphics. The process station provides a library of process elements and control function modules. These are selected and linked together via process lines and control signal lines as required. The second step is to enter process data and initial conditions. As more information about process and control becomes available, the simulation model will become more complete. The latest control database will be continually transferred to CADAS. The CADAS approach to simulation will ensure that the design team is fully conversant with the simulation model at all times. In projects where simulation is used for training or for verifying a process unit in the later stages of the design, the design team is either not involved in the simulation or does not have time to fully understand the model and participate in its building.

168 | INDUSTRIAL PROCESS CONTROL

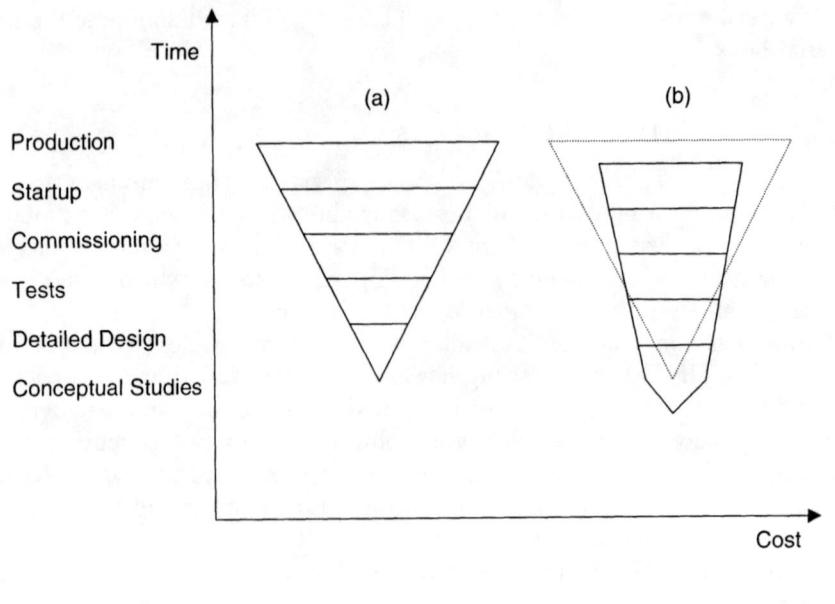

[Figure: Two inverted triangles comparing design methods. Y-axis labeled "Time" with levels: Production, Startup, Commissioning, Tests, Detailed Design, Conceptual Studies. X-axis labeled "Cost". (a) Orthodox design methods shows larger triangle; (b) Design based on CADAS shows smaller triangle within dotted outline of (a).]

(a) Orthodox design methods (b) Design based on CADAS

Figure 7-11 Top-down design comparisons.

7.8 CONCLUSION

In the previous sections we studied the main features of six popular control and safety systems. The structure, communications, main system components, and their application aspects were briefly described. For more detailed information and the latest system description, readers should consult the vendors' documentation; however, the presentation here provides a basic understanding of the covered systems and a comparison among system applications. Each system has some specific features that makes it more suitable for particular applications. Those aspects of a system that are critical when we consider it for a particular project are as follows:

- System structure
- Compliance with relevant standards and codes of practice (OSI, TUV, SINTEF, etc.)
- Operator station and control unit capabilities (number of displays, number of I/O, scan/update time, availability of neural networks, etc.)
- Vendor experience
- Vendor size
- Vendor presence
- System costs (initial and long term, maintenance, spare, and expansion)

In the description of each system, I endeavored to cover these subjects, although some items (e.g., vendor presence) are self-evident. The previous generation of control and safety systems were application oriented, but some of the modern systems can fulfill the requirements of any process plant. Examples of applications that the previous generation of systems could not handle because they needed hardware/software from another vendor are the following:

- ESD, F&G
- Surge control
- Vessel control
- Power management
- Metering

I must emphasize that even some of the modern systems cannot meet these requirements at present. Such vendors are advised to upgrade their systems before their competitors capture a larger portion of the market.

There are still some applications that none of the current systems can handle. These may require very fast scanning (e.g., 1,000 per second or higher for condition monitoring) or special software (e.g., pattern recognition, neural networks). Those vendors who take advantage of the latest hardware/software developments and learn the benefits of systems theory will survive and prosper. Clients who choose an unsuitable system will suffer. Such systems may need a costly upgrade shortly after the startup, may cause frequent shutdowns, reduce production and quality, increase the cost of spare parts management and maintenance, and provide a poor operator interface. In my opinion, operator interface is by far the most critical aspect of any process control system.

Glossary of Words and Abbreviations

ABB Master	Asea Brown Boveri safety and automation system; Section 7.2
AC	Alternating current
A/D	Analog to digital (conversion)
AEC	Advanced extended controller; TDC 3000; Section 7.4.4
AIM	Simrad integrated control and safety system; Section 7.7
ALARP	As low as reasonably practicable
Aliasing	Folding of frequencies; Section 3.2.1
AM	Application module; TDC 3000; Section 7.4.4
AMC	Advanced multifunction controller; TDC 3000; Section 7.4.4
AMPL	ABB Master Piece Language; ABB Master; Section 7.2.7
API	Application programming interface
APM	Advanced process manager; TDC 3000; Section 7.4.4
APS	Abandon production ship
APT	Application productivity tool; SIMATIC PCS7; Section 7.5.7
ASCII	American standard code for information interchange
Bad PV	A process variable value outside acceptable limits
Baud	Unit of data transfer rate; normally 1 Bit per second
BC	Basic controller; TDC 3000, Section 7.4.4; Balancing Control, AIM, Section 7.7.4
BCL	Batch control language; SIMATIC PCS7; Section 7.5.6
BCS	Ballast control system; Section 2.8.3
BD	Blowdown
Bit	0 or 1; the smallest digital systems piece of information
Bridge	A device for transparently passing messages between two physical parts of the same network across a large geographical distance
Byte	8 Bits
C	Controller (function block) output, also a programming language
Cache	A fast access memory, used by CPU for intermediary storage of instructions

CAD	Computer-aided design
CADAS	Computer-aided design, analysis, and simulation; Simrad AIM; Section 7.7.7
C&I	Control and instrumentation
CCR	Central control room
CFC	Continuous function chart; AIM; Section 7.5.7
CHIP	Computer highway interface package; PROVOX; Section 7.3.1 and 2
CL	Control language; TDC 3000; Section 7.4.5
CM	Computer module; TDC 3000; Section 7.4.4
COM	Component object model
CPM	Critical path method
CPU	Central processing unit
CRT	Cathode ray tube
CSMA/CD	Carrier sense multiple access/collision detection
CU	Control unit; the module that provides control and data acquisition functions in control and safety systems
DC	Direct current
DCS	Distributed control system
DDC	Direct digital control
DH	Data highway
DPS	Dynamic positioning system; Section 2.8.2
EC	Extended controller; TDC 3000; Section 7.4.4
EPU	Electrical power unit; Subsea; Section 2.7.1
ESD	Emergency shutdown system; ESD1 & 2; TUV Class 6, 5, and 4; SIL 3 and 2; probability of failure on demand between 10^{-4} and 10^{-2}
Ethernet	A synchronous communications protocol (IEEE 802.3 Standard); medium control is carrier sense multiple access with collision detection (CSMA/CD); introduced by Xerox in 1976
EUC	Equipment under control
F&G	Fire and gas system; TUV Class 4, SIL 2, probability of failure on demand between 10^{-3} and 10^{-2}
FAT	Factory acceptance test
FB	Function block
FFBus	Foundation bieldbus; Section 2.5.3
Fieldbus	The communication system that will eventually replace the 4–20 mA and other individual cables between field instruments and the control system
FO	Fiber-optic

Glossary of Words and Abbreviations | 173

FPSO	Floating production, storage, and offloading vessels, which are used instead of fixed offshore oil and gas production platforms; Section 2.8
Gateway	A device for passing messages between two different networks (usually employing different protocols)
GBytes	Giga Bytes; 1,073,741,824 Bytes
GUI	Graphical user interface; SIMATIC PCS7; Section 7.5.4
H	High, alarm
Hart	Hiway addressable remote transducer (protocol developed by Rosemount in the mid 1980s)
HAZOP	Hazard and operability study
HH	High High, alarm
HIM	Hiway interface module; TDC 3000; Section 7.4.2
HIPPS	High-integrity process protection system; TUV Class 7/8; SIL 4, probability of failure on demand between 10^{-5} and 10^{-4}
HLPIU	High-level process interface unit; TDC 3000; Section 7.4.4
HM	History module; TDC 3000; Section 7.4.5
Host	Supervisory computer, server; Section 2.3.5
HPU	Hydraulic power unit; Section 2.7.1
HQ	Headquarters
HSE	High-speed Ethernet
HV	High voltage
HVAC	Heating, ventilation, and air conditioning system
H1, H2	Foundation fieldbus networks (low speed, high speed)
IEE	The Institution of Electrical Engineers
IEEE	The Institute of Electrical and Electronics Engineers
IFC	A control unit; PROVOX; Section 7.3.4
ILM	Interface link module
I/O	Input/output
I/P	Input
ISA	Instrument Society of America
ISO	International Standards Organization
IT	Information technology
K	PID controller gain, kilo (1,000, or 1,024 in memory units)
KBaud	1,000 Bits per second (normally)
KBits	1,000 Bits
Kbps	Kilo bits per second
KBytes	1,024 bytes
K_{ij}	Relative gain
K_p	Process gain

K_s	Controller gain
KWords	1,024 words
L	Low, alarm
LAN	Local area network
LL	Low Low, alarm
LC	Logic controller; TDC 3000; Section 7.4.4
LCD	Liquid crystal display
LCN	Local control network; TDC 3000; Section 7.4.2
LCR	Local control room
LEPIU	Low-energy process interface unit; TDC 3000; Section 7.4.4
LER	Local equipment room
LLPIU	Low-level process interface unit; TDC 3000; Section 7.4.4
LM	Logic manager; TDC 3000; Section 7.4.4
LV	Low voltage
mA	Milliampere
MAN	Metropolitan area network
MAP	Manufacturing automation protocol; SIMATIC PCS7; Section 7.5.3
MAS	Manufacturing automation system; TDC 3000; Section 7.4.2
MBaud	1,000,000 Bits per second (normally)
MBits	1,000,000 Bits
Mbps	Mega bits per second
MBytes	1,048,576 Bytes
MC	Multifunction controller; TDC 3000; Section 7.4.4
MCS	Master control station; topside control unit for subsea control systems; Section 2.7.3
MeOH	Methanol
MMS	Media management system; SIMATIC PCS7; Section 7.5.3
MODBUS	Gould Modicon protocol; an asynchronous protocol, used by serial links such as RS 232
MPFM	Multiphase flow meter; Section 2.6
MTBF	Mean time between failures
MTTR	Mean time to repair
MTU	Master terminal unit
MUX	Multiplexer
MWords	1,048,576 words
NFS	Network file system; SIMATIC PCS7; Section 7.5.3
NIU	Network interface unit
OBT	Optical bus terminal
OLE	Object linking and embedding; Section 2.4.2
OLM	Optical link module

O/P	Output
OPC	OLE for process control; Section 2.4.2
OS	Operator station
OSI	Open system interconnect
P&ID	Piping and instrument diagram; an important document for the configuration and development of an SAS database
PC	Personal computer
PCN	Process control network; Section 7.3.2
PCS	Process control system; that part of an SAS that handles the control and data acquisition, rather than safety
PFD	Process flow diagram; shows main process equipment and piping
PI	Proportional integral control
PID	Proportional integral derivative control
PIN	Plant interface network, PROVOX, Section 7.3.2; plant information network, TDC 3000, Section 7.4.2
PIU	Plant interface unit
PLC	Programmable logic controller; a control unit used for batch and sequence control; all modern SAS control units provide this function
PM	Process manager; TDC 3000, Section 7.4.4
PMS	Process management system
PO	Purchase order
Protocol	The rules and regulations used by network participants in order to establish connections, transfer data, regulate the use of the network, and avoid contention for resources. Protocols are either synchronous (e.g., Ethernet, Token Ring) or asynchronous (e.g., RS 232)
PROVOX	Fisher-Rosemount control system; Section 7.3
PS	Process station; Simrad AIM; Section 7.7.4
PSD	Process shutdown; SIL 1 and 0, probability of failure on demand between 10^{-2} and 10^{-1}
PV	Process variable
QA	Quality assurance
QC	Quality control
QMR	Quadruple modular redundancy
RAM	Random access memory; a memory that takes the CPU the same time to gain access to any part of it; it is also called direct memory, and the CPU has fast access to it because it is normally in the same board as the CPU (unlike cassette or floppy/hard disk)
ROC	Rate of change

ROM	Read-only memory
Router	A device for connecting several networks simultaneously and sending messages from one network to another
RS 232	A serial link communication standard, which normally uses MODBUS Protocol and for short distances (typically 10 to 20 meters)
RS 422/485	Similar to RS 232, but for long distances (typically up to 1,000 meters)
RTD	Resistance temperature detector
RTU	Remote terminal unit, also a version of MODBUS protocol
s	Laplace–transform operator
SAS	Safety and automation system
SAT	Site acceptance test
SCADA	Supervisory control and data acquisition
SFC	Sequential function charts; SIMATIC PCS7; Section 7.5.7
SIL	Safety integrity level; as defined by IEC 1508/1511/61508; Section 2.11
SINEC H1	SIMATIC system communication network; Section 7.5.3
SOE	Sequence of events
SP	Setpoint
SQL	Structural query language
T	Temperature
T/C	Thermocouple
TCP/IP	Transmission control protocol/Internet protocol
T_d	Derivative time; in PID control
TDC	Totally distributed control
TDC 2000	Honeywell's first generation of control systems; Section 7.4
TDC 3000	Honeywell control system; Section 7.4
T_i	Integral time; in PID control
TMR	Triple modular redundancy
Token Bus	A token-passing communications protocol (IEEE 802.4 Standard); based on MAP, introduced by General Motors
Token Ring	A token-passing communications protocol (IEEE 802.5 Standard); based on IBM's early token ring network
TUV	A certifying authority; Technischer Ubervanchungs Verein
UCN	Universal control network; TDC 3000; Section 7.4.2
UCP	Unit control panel
UCS	Unified control system; SENTROL; Section 7.6.1
UNIX	A widely used operating system in real-time systems (and business computers), developed at AT&T Bell Laboratories
UOC	Unit operations controller; PROVOX; Section 7.3.4
UPS	Uninterruptible power supply

US	Universal station; TDC 3000; Section 7.4.3
UWS	Universal work station; TDC 3000; Section 7.4.3
VCS	Vessel control system; Section 2.8
VDU	Visual display unit
VMS	Vessel management system; Section 2.8
WAN	Wide area network
Windows NT	A popular operating system, by Microsoft
Word	The number of bits that the CPU handles in one cycle
z	Z-transform operator
Δ	Difference
ΔT	Integral step length
1oo2	One out of two voting
1oo2D	One out of two voting with extensive diagnostics
2oo2	Two out of two voting
2oo3	Two out of three voting
2oo4	Two out of four voting
2oo4D	Two out of four voting with extensive diagnostics

References

1. Kalani, G. 1988. *Microprocessor Based Distributed Control Systems*. Englewood Cliffs, NJ: Prentice Hall.
2. Beale, E.M.L. 1968. *Mathematical Programming in Practice*. Pitman.
3. Fisher-Rosemount. 1997. *Fieldbus Technical Overview*.
4. Hodgkinson, Geoff. "Less is More." *IEE Review*. September 1998.
5. Schickhuber, Gerald, and Oliver McCarthy. "Distributed Fieldbus and Control Network Systems." *Computing & Control Engineering Journal*. February 1997.
6. "Multiphase Meters and Their Subsea Applications." *Subsea Engineering News*. April 14, 1993.
7. Life Cycle Simulation. CADAS, Simrad AS.
8. AIM Documents, Kongsberg Simrad AS, P.O. Box 483, N-3601 Kongsberg, Norway; Knut Johansen, tel. 32-285324.
9. Churchley, Andrew. "Reliability of Microprocessor Based Protection Systems." *Measurement and Control*. Volume 27, December-January 1994–1995.
10. Goble, William M. "Building Reliable Control Systems." *InTech*. November 1994.
11. Hellyer, F.G. "The Application of Reliability Engineering to High Integrity Plant Control Systems." *Measurement and Control*. Volume 18, June 1985.
12. PROVOX Documents, Fisher-Rosemount Limited, Meridian East, Leicester, England LE3 2WU; John Nicholas, tel. (01162)-822664.
13. Shannnon, C.E. "A Mathematical Theory of Communication." *Bell System Technical Journal*. Volume 27, July and October 1948.
14. Meinert, Greg. "Openness for Automation Networks." *InTech*. October 1995.
15. Stoneham, Dave, and Colin Dowsett. "Technological Advances in Open Systems for Real-Time." *Measurement and Control*. Volume 28, November 1995.
16. Duin, William ver. "Solving Manufacturing Problems with Neural Net." *Automation*. July 1990.
17. Montgomery, Edward R. "Simulator Speeds Operator Training." *InTech*. October 1994.
18. ABB Master Documents, ABB Process Automation Limited, Gunnels Wood Road, Stevenage, Herts, England SG1 2EL; John Wilkins, tel. (01438)-742368.
19. TDC 3000 Documents, Honeywell Control Systems Limited, Lovelace Road, Souther Industrial Area, Bracknell, Berks, England RD12 8WD; Stephen K. Day, tel. (01344)-656793.
20. Simatic PCS Documents, Siemens AS, Oestre Aker Vei 90, P.O. Box 10, Veitvet, N-0518 Oslo, Norway; Bjorn Egil Mork, tel. 22-633985.
21. Sentrol Documents, Silvertech Limited, Planet House, North Heath Lane, Horsham, West Sussex, England RH12 4QE; Steve Ward, tel. (01403)-211611.

Index

ABB Master, 137–144
   Master Bus 300, 138–140
   Master Bus 200, 140
   Master Bus 90, 141
   Master Electronic Weighing, 143
   Master External Interface, 143
   Master Fieldbus, 141
   Master Host, 143
   Master Piece Language, 143
   Master Speed Drive Control, 144
   Master View, 141–142
   Advant Controllers, 142
   Advant Station 500, 141
   Abbreviations, 171–177
   A/D conversion, 56–58
   Advanced control, 106–114
Algorithms, 106–118
   Lead/Lag, 117
   Multivariable, 111–114
   PID, 115–116
   PID Gap, 106–107
   PID error squared, 111
   Square root, 117
   Traditional, 114–118
Aliasing, 57–58
Application engineering of control systems, 105–136
Application software development, 96
Artificial intelligence, 119–121
Availability Analysis, 41–46

Ballast control, 39

CADAS, 167–168
Commissioning, 98–99
Condition monitoring, 40–41
Control algorithms, see *Algorithms*
Control loop tuning, 105–106
Control of an irregularly shaped tank by neural networks, 121–129
Control pods, see *Subsea control and instrumentation*

Control theory, 58–60
Control unit, 15–16
Critical path methods, 86–98

DDC, 62
DISCOS, 155
Display hierarchy, TDC 3000 Universal Station, 150–152
Displays, VDU, 16–17
Distillation tower simulation, 135–136
Documentation, 101–103
Dynamic feedforward control, 107–108
Dynamic positioning, 38–39

Environmental, meteorological, and platform monitoring, 40
Ethernet, 11, 65–67
Expert systems, 120–121

Fieldbus, 26–33
Field instrumentation, 22–23
Fisher PROVOX, 144–148
   Control units, 147
   Engineering tools, 148
   Software/computing facilities, 147
Foundation fieldbus, 26–33
Function blocks, 114
Fuzzy logic, 120

Generic algorithms, 121

Hardware variety, 69–70
Heat balancing control, 108
Heater simulation, 130–132
Hierarchical system configuration, 62–64
HIPPS, 50–51
Honeywell TDC 3000, 148–154
Host, 18, 63

Information theory, 60–62
Integrated safety and automation systems, 73–79

181

Intelligent sensors (based on neural networks), 128–129
Intelligent systems, 118–129
Interactive methods, 111–114

LAN, 9–11, 62–63, 65–67
Lead/lag algorithm, 117
Level control by PID gap, 106–107

Maintenance strategies, 41
    Corrective maintenance, 41
    Detective maintenance, 41
    Predictive maintenance, 41
    Preventive maintenance, 41
Management computer, 64
Master control station, see *Subsea control and instrumentation*
MOD 300, 138
Multiphase flow meters, 33–35
Multivariable control, 111–114

Neural networks, 121–129

Objective, 1–5
OPC, 19–22
Open system technology, 64–68
Operational research, 2–4
Operator console, 83–85
Operator interface design, 95
Operator station, 16–17, 63
Optimization, 2–4

P4000, 138
P&ID, 7–9
PH control by PID error squared, 111–112
Philosophies (system), 1–2, 87
PID algorithm, 115–116
Process interface, 15–16
Process interface design, 94
Profibus, 23–26
Project engineering of control systems, 81–103
Project manning, 99–101

Qualitative reasoning, 121

Regulations, 89–93
Reliability analysis, 41–46
Risk analysis, 47–49

Sampling theory, 56–58
Scan rate (scan time), 58
Serial links, 11–12, 14
Siemens SIMATIC PCS7, 154–159
    Control units, 157–159
    Engineering interface, 159
    Operator interface, 157
    System communications, 156–157
SIL, 47–51
Silvertech Sentrol, 160–164
Simrad AIM, 164–168
    Application packages, 167
    Control units, 166
    Internal communications, 164
    Operator interface, 164–166
Simulation, 129–136
Smart instruments, 22
Software robustness, 70–71
Square root algorithm, 117
Staff training, 95–96
Standards and codes of practice, 89–93
Subsea control and instrumentation, 34–38
    Control pods, 34–35
    Master control station, 37–38
    Umbilical, 37
System analysis, 87
System configuration (SAS), 9–10
System hardware, 83–85
System requirements, 87–89
System software, 85–86
System testing, 97–98
Systems theory, 53–71

Tank process simulation, 133
TDC 2000, 15, 148
TDC 3000, 148–154
    Application packages, 154
    Control units, 152–153
    External interfaces, 154
    Host computer, 153
Token bus, 11
Token ring, 11
Training, simulation and, 132–136

Umbilical, see *Subsea control and instrumentation*
Universal operator station, 68–69

Vendor selection, 93–94
Vessel control systems, 38–40

Printed in the United States
133603LV00002B/35/A